Dimensional Analysis of Food Processes

Modeling and Control of Food Processes Set

coordinated by
Jack Legrand and Gilles Trystram

Dimensional Analysis
of Food Processes

Guillaume Delaplace
Karine Loubière
Fabrice Ducept
Romain Jeantet

ELSEVIER

First published 2015 in Great Britain and the United States by ISTE Press Ltd and Elsevier Ltd
Translated and adapted from *Modélisation en Génie des Procédés par Analyse Dimensionnelle*
© Lavoisier 2014.

ISTE Press Ltd
27-37 St George's Road
London SW19 4EU
UK

www.iste.co.uk

Elsevier Ltd
The Boulevard, Langford Lane
Kidlington, Oxford, OX5 1GB
UK

www.elsevier.com

Notices

For information on all our publications visit our website at http://store.elsevier.com/

British Library Cataloguing-in-Publication Data
A CIP record for this book is available from the British Library
Library of Congress Cataloging in Publication Data
A catalog record for this book is available from the Library of Congress
ISBN 978-1-78548-040-9

Printed and bound in the UK and US

Contents

Foreword

Dimensional Analysis, a powerful and analytical tool, has not only been around for a long time, dating back to Galileo, Newton, etc., but it is also truly interdisciplinary in nature, cutting across almost all branches of engineering and sciences. It is routinely used for the consolidation of data (both experimental and numerical) by reducing the number of variables (an outstanding example is the development of the Moody diagram for pipe friction which is really independent of equipment scale and fluid properties) and for the design of experiments to reduce the number of tests required, and/or with a view to scale up/scale down a process. Finally, it also provides a sound framework for simplifying the well-known momentum and energy equations used in the realm of transport of momentum, energy and mass in continua. Indeed, a proper scaling of these equations enables the development of useful solution methodologies of the nonlinear Navier–Stokes equations such as the notion of the boundary layer introduced by Prandtl in 1904 or the inviscid flow approximation. In chemical engineering, the comparison between chemical and physical kinetics gave birth to very useful non-dimensional numbers, such as Damköhler, Hatta and Thiele numbers, which allow us to define limiting phenomena to take into consideration for modeling the matter transformation process. In spite of its overwhelming potential, many textbooks do not bring out the full potential of *Dimensional Analysis*. For instance, most (if not all) books simply state the familiar Buckingham π theorem and illustrate the process of obtaining the pertinent dimensionless parameters via one or two examples, without much reflection on the physical significance of these parameters and/or how to use them for the purpose of scaling up/scaling down a process. This may be aptly called *"blind dimensional analysis"*, as discussed in Chapter 2 of this book.

The main interest of this book is to revisit the dimensional analysis for knowledge and scaling-up tools. For several decades, the numerical approach, when the problem can be investigated from phenomenological equations, or by using experimental designs coupled with statistical analysis, principal component analysis, for example, when all the phenomena of a problem are not completely controlled, as is often the case in food processing applications. The problems become very complex when the rheological properties evolve with space and time during the process. These constitute one of the main aims of this book. One of the key steps of dimensional analysis is the establishment of the independent variables of the problem. This list needs to have some knowledge of the process. The reduction of the number of variables due to the dimensional analysis leads to a dramatic decrease in the number of experiments together with the generalization of the obtained data.

This book endeavors to bring out the full potential of *Dimensional Analysis* in a systematic and gradual manner, with a particular emphasis on the processes dealing with the complex materials of multiphase nature and/or consisting of microstructures including oil–water emulsions, aerated systems, suspensions, etc. Such materials are routinely encountered in chemical, food, pharmaceutical, personal and health-care products, and smart electronic materials related applications. What sets this class of materials apart from the simple (low molecular weight or single phase) substances such as air and water is that most of them exhibit complex rheological properties that are strongly influenced by the presence and/or the evolution of their microstructures which itself may depend on the processing conditions, thereby emphasizing the intimate three-way link between structure, properties and processing, a paradigm which is well known in the field of polymeric materials. This aspect demands what might be called the *material similitude* in addition to the familiar geometric, kinematic, dynamic and thermal similarities required for the scale-up of processes involving simple substances.

This book outlines a systematic approach to use the *Dimensional Analysis* to scale up/scale down such processes by introducing the concept of *chemical* or *material* similarity. It also presents a framework as to what functional forms of the property – structure relationships including viscosity–shear rate, density–temperature, surface tension–temperature, etc. – are admissible which ensure material *similarity* (Chapters 4 and 5). Finally, Chapter 6, which is really the main chapter of the book, presents in detail six case studies drawn from diverse fields and arising from the personal engagement of the authors with wide-ranging industrial settings. These examples clearly demonstrate why it is seldom possible to achieve complete similarity between the lab-scale and full-scale operations. This difficulty arises from the unrealistic physical properties and/or the impractical operating conditions required for achieving the complete similarity. This book provides some guidelines to tide over these difficulties in practice. Of course, each situation needs to be examined on a case-to-case basis because the answer may vary well from one situation to another and/or even for the same application depending on the scale-up criterion to be used.

All in all, this is a very useful and interesting book which provides a fresh perspective of the age-old technique of *Dimensional Analysis*.

Finally, we are grateful to the authors for giving us the opportunity to write this foreword. During their research activities, the authors have demonstrated the relevance of the dimensional analysis to give a real universality of their data obtained in very complex problems. They were able to adapt dimensional analysis theory for modeling food processing in which evolutive non-Newtonian fluids are encountered. To the readers, we may say this is a genuine attempt to place *Dimensional Analysis* in its proper place, especially in the context of complex structured materials. We warmly recommend it to students, researchers and practicing professionals to explore and use this technique in their work. Enjoy *Dimensional Analysis*.

Jack LEGRAND
Professor of Chemical Engineering
GEPEA – UMR CNRS 6144
University of Nantes

Raj CHHABRA
Professor of Chemical Engineering
Indian Institute of Technology
Kanpur, India

Introduction

The community of chemical engineering has made significant progress up to now in identifying the mechanisms which govern the transformations of matter, especially agricultural and food materials. Indeed, the challenge of controlling food processes and making them reliable requires the modeling on different scales of the physicochemical and biological phenomena observed, with a view to:

– establishing the causal relationship which exists between the final characteristics of a product and the operating conditions under which the process takes place (geometry of the equipment, process parameters, physicochemical properties of the matter, initial and boundary conditions of the flow domain, etc.);

– and/or establishing the optimal performance of equipment from a range of possible settings, especially with regard to social and economic constraints (environmental regulations, eco-design process, economical costs, etc.).

Such modeling requires the resolution of a set of algebraic differential equations describing the physics of the system. Over the last two decades, the development of numerical methods and tools has produced models of increasing mathematical complexity, leading to significant progress (in this regard, computational fluid dynamics or multi-objective optimizations can be cited). Nevertheless, these numerical approaches do not systematically provide a framework for realistic modeling with regard to the required time for computation and the difficulty of accurate simulation:

– due to the limits of knowledge-based models with regard to turbulence, to the multi-phase nature of flows, to the chemical or biological kinetics, to the rheological behavior of fluids, and even more so when the equipment has complex geometry (e.g. extruder) and when the unit operations considered are of a multi-functional nature (e.g. heat-exchanger reactors) or are irreducibly coupled (e.g. simultaneous transfer of momentum-heat-electricity in ohmic heating).

In this context, we are convinced that the semi-empirical correlations between dimensionless numbers, deduced from dimensional analysis using the theory of similarity and experiments in laboratory-scale equipment (also called model), are a unique tool to fill the gaps left by "theoretical" modeling. Indeed, they provide clear and concise modeling of the process helping to:

– understand which are the most significant physical phenomena controlling the process;

– and, consequently, to identify the key mechanisms which need to be fully mastered in order to control the process completely.

Paradoxically, it should be noted that what occupied the minds of some of the greatest names in science, Galileo (1564–1642), Newton (1643–1727), Fourier (1772–1837), Froude (1810–1879) and Reynolds (1842–1912) to name but a few, is now rarely presented in teaching as a generic and forward-looking tool, and is therefore seldom applied.

Moreover, it is regrettable to note that dimensional analysis has evolved so little since it was founded and first applied to chemical engineering. For instance, most semi-empirical correlations between dimensionless numbers do not include the spatiotemporal variations of the physical properties of products during the transformation process. Indeed, these correlations are seldom constructed by considering the evolution of physical properties within the equipment and, more generally, throughout its entire transformation. Such approximations mean that the state of the system being studied is defined by a truncated (and insufficient) number of dimensionless numbers. This leads to non-generic process relationships since they are unable to describe the specific evolutions of the system resulting from the variability of the material's physical properties. It is clear that this is an extremely limiting constraint, especially for agro-food process engineers constantly confronted with the problem of conversion of material within reactors. This case is even more acute given the fact that physicochemical analysis techniques provide an ever finer characterization of the evolutions of the physical properties of matter subjected to complex stresses (temperature, shearing-time-deformation, pressure, etc.).

We believe that this partly explains the growing indifference toward, and even denigration of, dimensional analysis. The objectives of this book are:

– to review in detail the theoretical framework which allows the principles of similarity theory to be respected in the case of processes using a material with constant or variable physical properties in the course of the transformation;

– to give guidelines to follow in this case, and the tools available to rigorously construct a semi-empirical correlation between dimensionless numbers.

This book is the result of the experience acquired during various applied research projects, where a modeling with dimensional analysis was used to obtain revealing results. Moreover, detailed exchanges with J. Pawlowski and M. Zlokarnik, and the deciphering of their pioneering work on how to take into account the variability of material physical properties in the theory of similarity, helped us to improve our own understanding and present the underlying theoretical framework. This book provides case studies which were resolved using this logic.

To achieve this objective, we attempted to establish and define, in the clearest and most instructional manner possible, the methodology to adopt in order to produce, for materials with constant or variable physical properties, a dimensional analysis coupled with experiments carried out with laboratory or pilot-scale equipment to identify a process relationship.

This book is intended for students of chemical engineering, from Bachelor's level to PhD level, and, of course, for engineers and scientists looking to benefit fully from their experiments with laboratory-scale equipment and pilots. It also endeavors to promote and develop chemical engineering, by offering reliable, robust and relevant tools:

– to better model transformations of matter and the interactions between product and processes using a concise and physical view of phenomena;

– for the sizing, diagnosis and control of product transformation processes, reverse engineering and scale up/down.

We chose first to explain the foundations and the basis of dimensional analysis for processes involving fluids (materials) with constant physical properties. Therefore, the first three chapters discuss the interest of this approach (Chapter 1), detail the methodology step by step (Chapter 2) and then offer tools and advice for its implementation (Chapter 3). For ease of reading, the mathematical details have been reduced as much as possible in the body of the text and instead are included in the appendices when deemed necessary.

Chapter 4 shows how to extend the theoretical framework to include the dimensional analysis of materials which do not have a constant physical property during the course of the process. We will demonstrate that it is necessary to add supplementary variables to the list of physical quantities influencing the target variable (also known as the variable of interest). These describe the variability of the material's physical property (e.g. a rheological law or the dependence of viscosity on the temperature). We also introduce the notion of *material similarity*, which can be used to evaluate whether a process relationship established for one fluid remains valid for other fluids.

Finally, we analyze the consequences of these elements on the dimensionless numbers which define the system and the resulting process relationships, in order to carry out an extrapolation process for other scales or products (Chapter 5).

In order to illustrate the method as widely as possible, and to supplement the examples described in the first five chapters, Chapter 6 presents a selection of examples as applied to food transformation processes emerging from our own research.

Dimensionally yours,

Guillaume DELAPLACE
Karine LOUBIÈRE
Fabrice DUCEPT
Romain JEANTET

Objectives and Value
of Dimensional Analysis

The phenomena involved in matter transformation can be described by fundamental momentum, mass and energy transport equations, coupled with equations of chemical or biological kinetics and the constitutive and rheological equations. Boundary conditions of the flow domain and initial conditions are associated with this system of equations. Unfortunately, it is usually impossible to resolve this system of equations because:

1) the form of the differential equations is too complex to be integrated over the entire flow domain, especially given the complex geometry of the industrial equipment;

2) the numerical models used in order to reach an approximate resolution are often incomplete and/or insufficiently accurate to exhaustively describe the physics of phenomena (particularly turbulence, as well as coupled transfers, interactions between phases, rheological laws, etc.) and/or often too expensive to develop in terms of both and resource needed time.

For these reasons, scientists are still ill-equipped when facing the question of predicting the evolution of matter contained in equipment or quantifying the causal relationship which exists between the operating conditions imposed and the system's output variables (e.g. properties of the products or chemical conversion). In order to overcome such difficulties, experimental studies are, therefore, essential. They are either carried out on real industrial equipment (1/1 scale) or on a laboratory or pilot-scale (equipment representing the industrial equipment, generally reduced in size but geometrically similar). Carrying out experiments on laboratory-scale equipment is justified given the difficulty of experimenting on the real system (investment costs of a 1/1 scale installation, high experimentation costs given the volume of products to test and resources to mobilize, coupled with the considerations on hazards, etc.).

These experimental studies provide a first analysis of the empirical relationships linking the variable being studied (*target variable*) to a set of dimensional physical quantities. Such an analysis is not fully satisfying in terms of understanding the conditions to adopt in order to reproduce the transformation on a different scale (scale-up or scale-down). Indeed, the mechanisms governing the transformation process are dependent from the experiment scale on which the transformation has been carried out, as explained below (see section 1.2). Therefore, a given transformation mechanism is the product of a collection of operating conditions, notably including the scale of the experiment, and depending on the size of the system studied, the nature of the phenomena which occur may change. The operator is also ill-equipped to understand whether a different experiment scale will conserve all the phenomena between the laboratory-scale and the 1/1 scale.

To do this, dimensional analysis must be used to group the dimensional variables which influence the operation in the form of a set of dimensionless numbers characteristic of the state of the system, generally along with a precise physical meaning (see section 1.1 on the *configuration of the system*). It then becomes possible to determine which operating conditions should be chosen to obtain a phenomenological identity between the laboratory-scale and industrial equipments. To do so, the equality of the numerical value of each dimensionless number must be verified on both scales (conservation of an identical *operating point* at both scales). It is essential to understand that complying with this principle (known as *the theory of similarity*) is the only way to generalize the results obtained from one scale to another. This fact explains why the notions of model (laboratory-scale equipment) and similarity are closely linked and inseparable. This aspect is addressed in Chapter 5.

The key challenge for the researcher, therefore, lies in the ability to construct dimensionless numbers characteristic of the process being studied, and in the calculation of the set of numerical values of these dimensionless numbers, which define the operating points of the system. In this context, dimensional analysis provides a powerful theoretical framework required to generate an unbiased set of dimensionless numbers.

The construction of dimensionless numbers is traditionally carried out using dimensionless differential equations which:

– govern transport phenomena (momentum, mass and energy);

– define the boundary conditions of the flow domain and the initial conditions.

When the problem is too complex (geometry, formulation of the boundary conditions or geometry of the system, choice of the macroparameters used as target variables, etc.), another alternative consists of obtaining these numbers using the list of dimensional physical quantities influencing the *target variable*. In chemical engineering, this makes up the majority of cases. In this approach, traditionally called *"blind" dimensional analysis*, this list can be established without a complete level of physical knowledge on the process studied. In some cases, this may be very weak, and very advanced in other cases, depending on the scientific maturity of the process or the product formulated. Subsequently, the dimensionless numbers formed will have a more or less explicit physical sense.

In addition to the advantage of being able to be carried out without in-depth knowledge of the process, the "blind" dimensional analysis, coupled with the theory of similarity and experiments in laboratory-scale equipment, provides a global modeling of the process in the form of semi-empirical correlations (linking the target dimensionless number to a set of dimensionless numbers and translating the causes of the target variable's variation), which can be extrapolated to another scale or another product. In this book, these will be called *process relationships*.

Regardless of the approach used, the procedure for determining these dimensionless numbers characteristic of the process being studied is accompanied by a reduction in the number of variables by which the physical phenomenon[1] is described. This consequently minimizes the experiments to be undertaken to establish the process relationship linking the various dimensionless numbers (see section 1.3).

Furthermore, the process relationship established using the dimensional analysis coupled with the theory of similarity and experiments on laboratory-scale equipment helps to identify the predominant physical phenomena.

---

1 In other words, a phenomenon described by a set of m dimensional physical quantities expressed by n_d fundamental dimensions can then be described by a set of $(m-n_d)$ dimensionless numbers (see Chapter 2, section 2.1.3.3).

These are the phenomena which control the target variable insofar as it is based on dimensionless numbers which have a precise physical meaning.

Finally, this physical description coupled with the space reduction of the quantities describing the phenomenon can provide a simplified graphic representation. This, therefore, constitutes:

– an aid for understanding and carrying out the process;

– a way to carry out a reverse engineering approach (process for defining the set of operating conditions providing a given value of the target variable (see sections 1.3 and 1.4)).

This chapter illustrate the objectives and power of dimensional analysis.

1.1. Grouping dimensional variables in the form of a set of dimensionless numbers with a precise physical sense

The work of British physicist Osborne Reynolds (1842–1912) deals notably with the study and the characterization of liquid flows in cylindrical ducts. In 1883, by adding a dye, he illustrated the existence of various flow regimes for liquid flows in transparent pipes with a cylindrical section imposing various experimental conditions (average fluid velocity v, diameter of the pipe D, dynamic viscosity and density of the fluid μ and ρ, respectively). He noted that:

– the end of the laminar regime, shown by the appearance of vortices (displayed by the radial mixing of the dye), can be associated with a certain value of the dimensionless ratio:

$$\frac{\rho \cdot v \cdot D}{\mu} \qquad [1.1]$$

– this unique dimensionless number groups the set of dimensional variables, which influence the flow, and constitutes thus the configuration of the system.

As described in Appendix 2, this number, now known as the Reynolds number (Re), represents the ratio of two momentum fluxes (or forces) defined according to:

– the convective flux of momentum:

$$\varphi_i = \rho \cdot v^2 \cdot L^2 \tag{1.2}$$

where ρ is the density of the fluid, v is the average velocity and L is the characteristic length of the system being studied (here, the diameter of the pipe is D).

– the force linked to molecular momentum flux or the viscous effect, which constitutes a resistance to flow:

$$\varphi_j = \mu \cdot v \cdot L \tag{1.3}$$

Equations [1.2] and [1.3] give:

$$Re = \frac{\varphi_i}{\varphi_j} = \frac{\rho \cdot v^2 \cdot L^2}{\mu \cdot v \cdot L} = \frac{\rho \cdot v \cdot L}{\mu} \tag{1.4}$$

The Reynolds number, comparing the relative importance of these two characteristic forces in the flow, thus provides a physical analysis of the phenomena observed. In general, it is noted that the turbulent regime is reached around 4000, after which the level of turbulence continues to grow as the value of Re increases. Consequently, high values of Re show the dominance of the convection force compared to the viscous friction force. On the contrary, small values of Re show the predominance of the viscous friction force compared to the convection force and correspond to the laminar regime.

It should be noted that the Reynolds number can also be obtained by making the momentum transport equation dimensionless.

1.2. Constructing generic models which can be used on other scales

As mentioned in the introduction, mechanisms at the origin of the elementary processes which influence the performance of a process depend on the scale of the experiment. Consequently, to be able to extrapolate the results of the laboratory-scale equipment to a 1/1 scale (or vice versa), it is necessary to determine the conditions which must be fulfilled in order to guarantee the conservation of operating points at different scales.

This can be illustrated through a very simple example. Take two straight and smooth cylindrical pipes with different diameters which carry the same

Newtonian fluid at the same average velocity v. Depending on the diameter D, the layers of the fluid may slide past each other (laminar regime) or flow in the form of vortices (turbulent regime). It is not possible to say beforehand what the flow regime will be considering the dimensional variables separately v, D, μ and ρ. To have a clear idea of the flow regime, it is necessary to estimate the value of the Reynolds number Re, which establishes the system's configuration.

Subsequently, if the value of the Reynolds number is the same for the experiments on two different scales, the flow regime will be the same in the two cases. If this value is different, the flow regimes will be different, and in this case nothing guarantees that the phenomena linked to the flow (e.g. a pressure loss coefficient) will be identical on both scales. For example, consider a smooth, straight and cylindrical industrial pipe, with a diameter of $D_1 = 0.3$ m (scale 1), carrying water (at 20°C, kinematic viscosity $v = 10^{-6}$ m^2.s^{-1}) at a volumetric flow rate Q_1 of 30 m^3.h^{-1}. The average velocity of the fluid, v_1, is therefore equal to:

$$v_1 = Q_1 \bigg/ \left(\frac{\pi.D_1^2}{4} \right) = 0.118 \text{ m.s}^{-1} \qquad [1.5]$$

Now consider the flow in a laboratory pipe of reduced size with a scale factor (relationship between the industrial scale and study scale) $F_e = D_1/D_2$ equal to 30. The laboratory-scale equipment, therefore, has a diameter of $D_2 = 0.01$ m. The challenge is then to choose the volumetric flow rate or the average velocity of the fluid at the laboratory equipment scale in order to guarantee the same flow regime as in the pipe with a diameter of 0.3 m.

At scale 1 (on industrial equipment), the value of the Reynolds number is as follows:

$$Re_1 = \frac{v_1.D_1}{v} = \frac{0.118 \times 0.3}{10^{-6}} = 35\,400 \qquad [1.6]$$

The flow is, therefore, turbulent. If experiments are carried out on the laboratory-scale equipment at the same average velocity as for the experiment on the industrial equipment, this gives:

$$Re_2 = \frac{v_1.D_2}{v} = \frac{0.118 \times 0.01}{10^{-6}} = 1180 \qquad [1.7]$$

In this case, the flow regime is laminar. The phenomena observed with the laboratory-scale equipment would, therefore, be different from those present in the industrial equipment.

If the volumetric flow rate Q_I is maintained for the laboratory-scale equipment, an average velocity v_2 of 106 m.s^{-1} is obtained, which, as well as being difficult to implement in practice, leads to significantly more turbulent flows than on the industrial scale:

$$Re_2 = \frac{v_2.D_2}{\upsilon} = \frac{\left(Q_1 / \left(\frac{\pi.D_2^2}{4}\right)\right).D_2}{\upsilon} = \frac{106 \times 0.01}{10^{-6}} = 1\ 060\ 000 \qquad [1.8]$$

Using dimensional analysis and the theory of similarity, it can be demonstrated that in order to conserve the phenomena from one scale to another (and thus to conduct a rigorous extrapolation process), it is necessary to conserve the value of the Reynolds number. Therefore, to maintain $Re_I = 35400$ at both scales:

– if the same fluid is used (case number 1), it is necessary to adjust the average velocity in the laboratory-scale equipment according to $v_2 = 3.54$ m.s^{-1}, which corresponds to a volumetric flow rate of 1 m^3.h^{-1}, given:

$$Re_1 = Re_2 \qquad [1.9]$$

equals $\dfrac{v_1.D_1}{\upsilon} = \dfrac{v_2.D_2}{\upsilon}$ \qquad [1.10]

The conservation of Re (equation [1.10]) then makes the relationship of the velocities and scale factors equal to:

$$\frac{v_2}{v_1} = \frac{D_1}{D_2} = F_e \text{ hence } v_2 = 0.118 \times 30 = 3.54 \text{ m.s}^{-1} \qquad [1.11]$$

– if we conserve the same average velocity $v_I = 0.118$ m.s^{-1} for both scales (case number 2), it is necessary to change the fluid. In this case, experiments on the laboratory-scale equipment may be extrapolated to an industrial scale only if the fluid used in the laboratory has a kinematic viscosity of $\upsilon_2 = 3.33 \times 10^{-8}$ m^{2}s^{-1}, which is possible with certain gases. Indeed:

$$Re_1 = Re_2 \qquad\qquad [1.12]$$

equals $\dfrac{v_1 \cdot D_1}{\upsilon_1} = \dfrac{v_1 \cdot D_2}{\upsilon_2}$ \qquad\qquad [1.13]

hence $\upsilon_2 = \upsilon_1 \cdot \dfrac{D_2}{D_1} = \dfrac{\upsilon_1}{F_e} = 3.33 \times 10^{-8} \ \mathrm{m^2.s^{-1}}$ \qquad\qquad [1.14]

1.3. Reduce the number of experiments by providing a synthetic and physical view of the phenomena

According to the Vaschy–Buckingham theorem, any physical quantity representing a phenomenon (any target variable) function of m physical quantities measured by n_d fundamental dimensions can be described by an implicit function between m-n_d dimensionless numbers. Consequently, the number of experiments to undertake in order to establish the process relationship linking the various dimensionless numbers is significantly reduced[2].

To illustrate this, take the example of the linear pressure drop ΔP_L (pressure loss per unit of length, expressed in Pa.m⁻¹), for the flow of a Newtonian fluid in a straight, smooth and cylindrical pipe, when the regime is fully established (no entrance effects).

A physical analysis in this case provides a list of physical quantities which have an impact on the variable being studied: the average velocity v, the characteristics of the fluid, density ρ and kinematic viscosity υ, and the pipe diameter D. The linear pressure drop, therefore, depends on four physical quantities. To experimentally analyze the influence of each of these physical quantities on the linear pressure drop, it is necessary to vary one of the quantities while keeping the others constant. If each of these takes six values successively, it is necessary to carry out 6^4 tests, or 1296 experiments. Apart from being a time-consuming process, it is also

---

2 Of course, this property does not prevent the process being applied to a large number of experimental data previously obtained or readily available in other writings on the subject.

difficult to carry out a general analysis of the results obtained, and to gain a generic understanding of the physics of these phenomena.

It can be demonstrated that the dimensional analysis of this problem leads to a relationship F between the two relevant dimensionless numbers:

$$\frac{\Delta P_L.D}{\rho.v^2} = F\left(\frac{\rho.v.D}{\upsilon}\right)$$ [1.15]

The "target" dimensionless number which characterizes the linear pressure drop (left-hand term in equation [1.15]) is known as the friction factor, which is solely dependent on the Reynolds number. The physical significance of the friction factor is described in Appendix 2. Disregarding the numerical factors, this is the ratio of:

– the friction of the fluid against the wall

$$\varphi_i = \tau_p \cdot L^2$$ [1.16]

where τ_p is the wall shear stress, with $\tau_p = \Delta P_L.D/4$;

– and the inertial force:

$$\varphi_j = \rho \cdot v^2 \cdot L^2$$ [1.17]

Thus, dimensional analysis provides an overall physical interpretation of the problem by showing that the friction factor is linked to the hydrodynamics in the pipe (the flow rate), and not to the individual effect of the dimensional quantities v, D and υ.

Moreover, the number of experiments required to model the relationship in [1.15] with the same accuracy as before will, therefore, only be (a minimum of) 6 instead of the 6^4 initial experiments. Using the available experimental results, it can be shown that the resulting graphic representation is given in Figure 1.1.

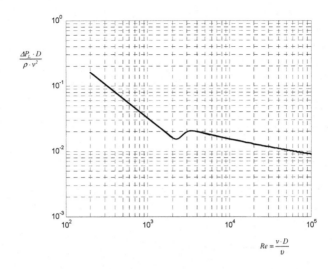

Figure 1.1. *Friction factor versus Reynolds number plot in a smooth duct*

1.4. A tool for carrying out processes and assisting a reverse engineering approach

Using the process relationship shown in Figure 1.1, it is possible to go back to the relationships linking the dimensional quantities and determine the parameter values of the process required to achieve the desired value of the target variable. It is, therefore, possible to generate a graph of pressure drops (Figure 1.2) identical to the representation available in technical documents. These graphs make it possible to quickly evaluate the linear pressure drop associated with the flow of a fluid of known properties in a duct of a given diameter. For instance, Figure 1.2(a) shows that the pressure drop associated with the flow of a fluid with a kinematic viscosity of 10^{-6} m^2.s^{-1} (water at 20°C) at a volumetric flow rate of 10 m^3.h^{-1} in a pipe with a diameter of 0.025 m is equal to 0.3 MPa for 100 m.

Conversely, implementing a reverse engineering approach, this representation helps determine the values which may be taken by the couple {Q, D} limiting the pressure drop at an acceptable level for a fluid of given kinematic viscosity. Therefore, it is possible to identify in Figure 1.2 four couples {Q ; D} (1.2 m^3.h^{-1} and 0.025 m, 10 m^3.h^{-1} and 0.05 m, 30 m^3.h^{-1} and 0.075 m, and 70 m^3.h^{-1} and 0.1 m) leading to a pressure drop value of 0.01 MPa for 100 m for water at 20°C.

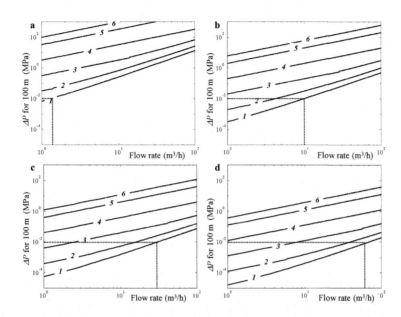

Figure 1.2. *Example of graphs showing the pressure drop for a 100 m duct values of depending on the flow rate in a smooth duct of given diameter (a): D = 0.025 m; b): D = 0.05 m; c): D = 0.075 m; d): D = 0.1 m) and six kinematic viscosities (1: $v = 10^{-6} m^2.s^{-1}$; 2: $v = 10^{-5} m^2.s^{-1}$; 3: $v = 10^{-4} m^2.s^{-1}$; 4: $v = 10^{-3} m^2.s^{-1}$; 5: $v = 10^{-2} m^2.s^{-1}$; 6: $v = 3.10^{-2} m^2.s^{-1}$)*

2

Dimensional Analysis: Principles and Methodology

This chapter examines the principles and methodology involved in dimensional analysis, which should be regarded in this context as a tool for establishing the dimensionless numbers linking the causes of the phenomena studied to its effects, using the homogeneity of dimensions.

The objective is to provide definitions, prerequisites, tools and essential mathematical demonstrations in order to both construct an unbiased space of dimensionless numbers associated with the phenomena studied and to illustrate the rigor and the potential of applicability of the dimensional analysis.

Dimensional analysis is a generic and multi-disciplinary approach: it will be applied in this book in order to understand the physics of, and thereby optimize, agro-food processes.

We strongly advise readers, whether trained or not, to take the time to consider these theoretical aspects before turning to the examples given in Chapter 6. This chapter addresses the semantics of the dimensional analysis process and provides several keys for understanding the rules (coherence of unit systems and independence of variables) and choices made (unit systems, repeated variables and rearrangements of dimensionless numbers) with which the user is faced in constructing the dimensionless numbers. Knowledge of all these aspects will allow the readers to feel confident in carrying out their own dimensional analysis modeling and/or in analyzing other works using this approach.

This chapter is divided into four sections:

– in section 2.1, basic concepts are defined and the fundamental rules for constructing the set of dimensionless numbers are provided;

– section 2.2 examines the elements which allow the readers to choose (1) a set of dimensionless numbers suitable for the experimental program implemented, and (2) the mathematical form of the process relationship correlating the dimensionless numbers with each other;

– section 2.3 focuses on the definition of notions regarding the configuration of the system and operating points, which are basic elements for addressing problems of process scale-up or scale-down (Chapter 5);

– section 2.4 concludes with a guided example, illustrating all the notions and rules previously described.

2.1. Terminology and theoretical elements

2.1.1. *Physical quantities and measures. Dimensions and unit systems*

2.1.1.1. *Physical quantities and dimensions*

Science uses physical quantities (or entities)[1] which need to be measured or calculated in order to establish predictive mathematical relationships linking the causes of a studied phenomenon to its effects. Such quantities may be a force, a moment, a mass and a velocity of bodies (solid, liquid or gas), energy or time, etc.

Each quantity has a number of independent properties: it's scalar or vector nature, its intensity, sign and dimension.

For instance, if the quantity is a velocity, it has a vector nature and its dimension is a length per unit time. However, if the quantity being studied is a distance, it has a scalar nature and its dimension is a length.

---

1 In the literature ([SZI 07] being a notable example), a classification of physical quantities can be found according to *constants* (physical quantity which never varies, e.g. the velocity of light in a vacuum or Planck's constant), *parameters* (physical quantity which is constant in the context in which it appears, e.g. acceleration of gravity) and *variables* (physical quantities which are varied during the study). This distinction will not be made in this book.

Therefore, it will be noted that physical quantities which describe a phenomenon have several properties and that one of them is their *dimension*, which specifies the physical nature of the quantity.

2.1.1.2. *Measure of physical quantity*

Several possibilities are available to the user for expressing the external or internal measure of a physical quantity.

2.1.1.2.1. External measure

In order to obtain the *external measure* of a physical quantity, its dimension is referenced in terms of a *standard* outside the system. This is why the numerical value of a physical quantity is expressed in the form of a number (*measure*) and a unit (*external standard*):

$$\text{numerical value of a physical quantity} = \text{measure} \times \text{unit} \qquad [2.1]$$

Therefore, the measure of a physical quantity cannot be unique and depends on the reference unit (the standard).

EXAMPLE 2.1.– External measure (I)

When the dimension is a length, the meter can be used as a standard. 0.2 is, therefore, the measure of the physical quantity which corresponds with a double decimeter. This physical quantity has length as its dimension is expressed in meters (unit).

However, regardless of the unit chosen, the dimension (length in this case) of physical quantity (distance in this case) is always identical and its "absolute intensity" is unchanged. By specifying the unit, the dimension (length) is provided, as well as the standard (meter), which is taken as a reference. Thus, quantities can be compared with each other.

The measure of the physical quantity obtained is considered as being *external* because the reference taken to standardize the physical quantity is not part of the system being studied (in example 2.1, the meter is defined by a reference length which is not part of the system being studied).

In order to express this external measure, a dimensional system consisting of *fundamental or basic dimensions* must be chosen. There are various dimensional systems in use: they can be classified according to whether they use a force or a mass as a fundamental dimension, or whether they are metric or non-metric. For instance, systems such as centimeter-gram-second (CGS), meter-kilogram-second (MKS) or American/British (foot/inch-pound-second) can be cited.

The system officially adopted by most countries and highly recommended is the *International System of Units (SI)*. It is based on *seven fundamental dimensions*, which are independent and sufficient for expressing the dimensions of all quantities encountered in the physical world. These fundamental dimensions, and the symbols chosen in this book to represent them, are shown in Table 2.1.

The universal definitions of the standards associated with each fundamental dimension in the SI were adopted at the 11th General Conference on Weights and Measures in 1960. For instance, the meter, unit of length, is defined as the length of path traveled by light in vacuum during a time interval of $1/299,792,458$ of 1 s.

It should be noted that the SI is also known as MKS (for meter, kilogram and second). The symbols for the units associated with the fundamental dimensions are given in Table 2.1.

The advantage of fixing the definition of units (standards) is that it avoids any error with regard to the external measure, and therefore allows comparisons between physical quantities.

Fundamental dimension: name and symbol		Associated SI unit: name and symbol	
Length	L	Meter	m
Mass	M	Kilogram	kg
Time	T	Second	s
Temperature	K	Kelvin	K
Quantity of matter	N	Mole	mol
Intensity of electrical current	I	Ampere	A
Light intensity	I_v	Candela	cd

Table 2.1. *Fundamental dimensions in the International System of Units (SI)*

It should be noted that even though the SI is highly recommended, *the choice of units is completely arbitrary*. For instance, Chapter 3 shows that it is possible to split a fundamental dimension into several fundamental dimensions of the same nature, or to introduce a new fundamental dimension.

In most cases, the unit of a physical quantity is not a fundamental unit, but is formed as a product of powers of several fundamental units derived from algebraic relationships linking the various physical quantities. In this case, they are called the *derived or secondary units*.

In terms of notation, it is common to use square brackets so as to make the dimension of a physical quantity explicit.

EXAMPLE 2.2.– Derived unit

The unit of velocity, v, is a derived unit, representing the distance traveled per unit time. Therefore, the dimension of a velocity is:

$$v \;\rightarrow\; [L.T^{-1}]v \rightarrow [L.T^{-1}]v \rightarrow [L.T^{-1}]v \rightarrow [L.T^{-1}]v \rightarrow [L.T^{-1}] \qquad [2.2]$$

It is expressed in m.s^{-1} in the SI.

A list of the most commonly used derived physical quantities and their dimensions in the SI system is shown in Table 2.2.

Physical quantity	Dimension
Area	$[L^2]$
Volume	$[L^3]$
Rotation speed, shear rate and frequency	$[T^{-1}]$
Velocity and conductance	$[L.T^{-1}]$
Acceleration	$[L.T^{-2}]$
Kinematic viscosity and diffusivity	$[L^2.T^{-1}]$
Density	$[M.L^{-3}]$
Surface tension	$[M.T^{-2}]$
Dynamic viscosity	$[M.L^{-1}.T^{-1}]$
Momentum	$[M.L.T^{-1}]$

Force	$[M.L.T^{-2}]$
Pressure and tension	$[M.L^{-1}.T^{-2}]$
Mechanical energy, work and torque	$[M.L^2.T^{-2}]$
Power	$[M.L^2.T^{-3}]$
Specific heat	$[L^2.T^{-2}.K^{-1}]$
Thermal conductivity	$[M.L.T^{-3}.K^{-1}]$
Heat transfer coefficient	$[M.T^{-3}.K^{-1}]$
Illuminance	$[cd.L^{-2}]$
Voltage	$[M.L^2.T^{-3}.A^{-1}]$
Molar mass	$[M.n^{-1}]$

Table 2.2. *Some derived quantities and their dimensions in the International System of Units*

Nevertheless, some derived units in the SI are named after well known scientists.

EXAMPLE 2.3.– Units named after scientists

The *newton* (N) is the unit of force in the SI; in terms of fundamental units, it is expressed as $1\ N = 1\ m.kg.s^{-2}$.

The *watt* (W) is the unit used for power. It also corresponds to a product of fundamental dimensions:

$$\text{power} = \frac{\text{energy}}{\text{time}} = \frac{\text{force} \times \text{length}}{\text{time}} = \frac{\text{mass} \times \text{acceleration} \times \text{length}}{\text{time}}$$

$$\text{hence its dimension is} : \left[\frac{M.L.T^{-2}.L}{T}\right] = \left[M.L^2.T^{-3}\right] \qquad [2.3]$$

Other examples are the unit for energy or work, the *joule* (equivalent to $kg.m^2.s^{-2}$); the unit of electrical resistance, the *ohm* (equivalent to $kg.m^2.s^{-3}.A^{-2}$); the unit of pressure, the *pascal* (equivalent to $kg.m^{-1}.s^{-2}$) and so on.

Some *derived quantities* can also be *dimensionless*, such as:

– *plane and solid angles*, respectively, expressed in radians (rad) and steradians (sr). Indeed, any plane angle is defined as the ratio between two lengths: the length of an arc and the radius of the associated circle. A solid angle (which can be considered as a three-dimensional angle at the top of a cone coinciding with the center of a sphere) is defined in the same way as the ratio between two areas: the area of a segment of a sphere and the radius of this sphere squared;

– *volume or mass fractions* which correspond, respectively, to the ratios of two volumes or two masses;

– *shape factors* (circularity, sphericity, etc.).

From this, it can be concluded that, in order to constitute a system of measures, it is necessary:

– first, to select the fundamental dimensions and their numbers, as well as the units used to measure them;

– second, to define the derived quantities.

It will become clear later that this second point is essential in order to fulfill the principle of homogeneity.

Only the SI has been presented here, but as stated previously, there are several systems (e.g. CGS), which have their own fundamental dimensions. It should be noted at this stage that these choices are arbitrary. On the basis of the previous considerations, there is no reason why the SI should be preferred over any other system to express the dimensions of physical quantities. *The only obligation for correctly carrying out dimensional analysis is for the system of units (fundamental and derived units) to be coherent.* To achieve this, the following condition must be respected: *the numerical coefficients of proportionality which appear when derived units are deduced from base units (using formulae which translate physical laws) must be equal to 1.* For example:

– the meter per second is the only coherent derived unit in the SI system. Indeed, the kilometer per hour and the centimeter per second are not coherent units in the SI system. There are respective conversion factors with the meter per second, respectively, equal to 1000/3600 and 0.01, and therefore are different from 1;

– even though the degree Celsius is commonly used, the fundamental unit of temperature in the SI is the Kelvin. The ambiguity comes from the fact that it is often temperature differences which are considered as physical quantities of interest, and in particular cases, using the degree Celsius or the Kelvin is equivalent:

$$t_2(^\circ C) - t_1(^\circ C) = T_2(K) - T_1(K)\theta_{\text{celsius 2}} \; (^\circ C) - \theta_{\text{celsius 1}} \; (^\circ C) = \theta_2(K) - \theta_1(K) \qquad [2.4]$$

For dimensional analysis modeling, the coherence of a given unit system ensures that a process relationship (as will be defined in this chapter) established in this system will remain valid if another coherent unit system is used.

All dimensional analysis are carried out in this book using the SI. This coherent system has been adopted in most countries around the world and is also the system of choice for the scientific community for comparing external measures of physical quantities (orders of magnitude).

2.1.1.2.2. Internal measure

We have shown that the first possibility available to the user in terms of expressing the measure of a physical quantity is its external measure. The second possibility is to express the physical quantity using *an internal measure*[2]. To do so, the value of the physical quantity in question is referenced in terms of a set of physical quantities belonging to the studied system, so as to constitute a dimensionless number. The set of quantities used to make the physical quantity dimensionless is called *repeated physical variables* and will constitute a base.

EXAMPLE 2.4.– Internal measure (I)

The Reynolds number is defined by:

$$Re = \frac{\rho.v.D}{\mu} \qquad [2.5]$$

which characterizes the flow of a Newtonian fluid (of density ρ and dynamic viscosity μ) flowing with an average velocity v in a pipe of diameter D.

---

2 This notion was first introduced by László [LÁS 64].

It can be seen as an internal measure of viscosity μ when repeated physical variables are the set (ρ, v and D). If the repeated physical variables chosen are (ρ, μ and D), the Reynolds number Re represents an internal measure of velocity v.

It should be noted that the isolated observation of an internal measure does not allow all of the repeated physical variables used as standards to be identified or to differentiate them from others (known as non-repeated physical variables). It is only through the observation of several internal measures characterizing the physical process studied (before the rearrangement of internal measures[3]) that we might distinguish the repeated variables from the others.

When this notion of internal measure is generalized (i.e. when we select and use some physical quantities of the system studied as standards), the dimensionless numbers are obtained for all the physical quantities other than the repeated variables. We will use this technique a little later to build the internal measures characterizing any physical phenomenon of interest.

2.1.2. The principle of homogeneity

Dimensional analysis is based on a simple physical principle:

The mathematical formulation proposed to describe a physical phenomenon must be dimensionally homogeneous.

This means that the physical relationship which links the physical quantity V_1 to other physical quantities V_2, V_3, V_4 and V_5 where:

$$V_1 = V_2.V_3 + \frac{V_4}{V_5}$$ [2.6]

only makes sense if $V_2.V_3$ and V_4/V_5 have the same dimension as V_1.

It is essential to remember that this principle of homogeneity does not apply to quantities which have not been rigorously defined by a mathematical relationship. For instance, this is the case for thermal sensations characterized in the Bedford scale by a number ranging from 1 to 7 for physiological

---

3 This is explained further in section 2.1.3.5.

impressions ranging from "too hot" to "too cold". This is clearly explained by the fact that dimensional analysis assumes to have previously set out an appropriate system of units, notably ensuring that all the physical quantities other than the fundamentals are linked to it by algebraic relationships. This condition is not guaranteed in the case of the Bedford scale.

It immediately becomes clear that the use of the principle of homogeneity shows whether or not a mathematical relationship with dimensional physical quantities is correct. This is a process which we highly recommend to our students.

However, this property of dimensional homogeneity of equations has many other consequences. Above all, the dimensional analysis of physical quantities constituting an algebraic relationship helps construct an equivalent system of dimensionless numbers. This dimensionless formulation, known as dimensional analysis, provides the key advantage to describe the physical phenomenon using a set of internal measures whose number is smaller than the initial number of physical quantities. The application of the Vaschy–Buckingham theorem, also called the π -theorem, helps us to determine the *maximum*[4] number of internal measures needed to describe the physical phenomena studied (see section 2.1.3.3).

2.1.3. *Fundamental rules for constructing the set (or space) of dimensionless numbers associated with a physical phenomenon*

The methodology for describing a physical phenomenon by a set of dimensionless numbers has a number of steps:

1) choosing the target variable, identifying and establishing the list of relevant independent physical quantities which influence the target variable;

2) determining the dimensions of the physical quantities listed;

3) applying the Vaschy–Buckingham theorem to obtain the number of internal measures governing the phenomenon analyzed;

---

4 The work of Saint-Guilhem [SAI 62] has shown that the minimum number of internal measures is not immediately achieved, and that it can be reached by adapting the system of units to the problem being studied.

4) constructing dimensionless numbers (elimination/substitution method, matrix operations method and shift relationships between spaces of dimensionless numbers);

5) if applicable, rearranging the dimensionless numbers to produce internal measures with a more intuitive or suitable meaning with regard to the experimental program implemented to establish the process relationship.

2.1.3.1. Step 1: choosing the target variable and establishing the list of relevant independent physical quantities

First and foremost, the problem must be defined. To start, it is necessary to identify the *target variable* (also called the variable of interest) which is characteristic of the phenomenon being studied. This is a quantifiable indicator of the performance of the process, such as the mixing time or the size of droplets.

Subsequently, all the physical quantities influencing the phenomenon being studied are listed (the causes), without discriminating according to the degree of influence each is supposed to have. At this stage, we advise the readers to carry out the most exhaustive list possible.

Finally, *the physical independence of the physical quantities listed must be verified.* This means ensuring that no physical quantity, except the target variable, is affected by the variations of the other quantities. If this is the case, this physical quantity must be eliminated. For instance, in the study of mixing time in a stirred tank, it is not possible to simultaneously list the impeller rotational speed and the power required to drive the stirring device since these two physical quantities are two indicators of the same cause: the rotation of the impeller in the mixing system. These two indicators are linked to each other and cannot be considered as independent of each other. Even if the analytical relationship between these two variables is not exactly known, only one of these two physical quantities should be listed.

Establishing the relevance list of independent physical quantities is the most difficult step because there is no predefined rule. It is a matter of judgment, where the knowledge of the physical phenomena involved and experience have the greatest significance. Nevertheless, certain elements of the methodology are given in Chapter 3.

2.1.3.2. *Step 2: determining the dimensions of the physical quantities listed*

The dimension of each physical quantity listed is expressed in terms of fundamental dimensions. No condition other than the coherence of the system of units is required to express these dimensions. The number of fundamental dimensions depends on the measure system adopted and the complexity of the physical phenomenon studied.

It is then possible to construct a matrix, known as a *dimensional matrix* and denoted as **D**, where:

– the number of columns is equal to the number of physical quantities listed;

– the number of rows is equal to the number of fundamental dimensions required to express the dimension of the set of physical quantities listed.

The dimensional matrix **D** contains the exponents e_i^k (*k* refers to the fundamental dimension, and *i* refers to the physical quantity listed) to which it is necessary to individually raise each fundamental dimension (constituting the rows) so that the physical variable at the top of the column is homogeneous with a product of fundamental dimensions raised to the exponents in question.

This is illustrated in example 2.5, which we will continue to use in the various steps in the methodology.

EXAMPLE 2.5.– Fundamental rules for constructing the set of dimensionless numbers associated with a physical phenomenon: steps 1 and 2 (I)

Consider a completely imaginary example in which the variable of interest, denoted as V_1, is influenced by a set of five independent physical quantities $\{V_2, V_3, V_4, V_5, V_6\}$.

The number of fundamental dimensions has been set at 3. These dimensions are denoted here as d_k so that it can be generalized to any system of units being used. In the SI, they correspond to M, L, T, etc. (Table 2.1) and their units are kg, m, s, etc.

The dimension of each physical quantity is expressed in terms of fundamental dimensions in equation [2.7] (the numerical values of the exponents were randomly chosen):

$$
\left.\begin{aligned}
[V_1] &= d_1^0 \cdot d_2^{-2} \cdot d_3^1 \\
[V_2] &= d_1^0 \cdot d_2^{-1} \cdot d_3^0 \\
[V_3] &= d_1^1 \cdot d_2^{-3} \cdot d_3^2 \\
[V_4] &= d_1^1 \cdot d_2^0 \cdot d_3^{-3} \\
[V_5] &= d_1^1 \cdot d_2^{-1} \cdot d_3^{-1} \\
[V_6] &= d_1^2 \cdot d_2^1 \cdot d_3^0
\end{aligned}\right\} \Leftrightarrow \qquad [V_i] = \prod_{1 \le k \le 3} d_k^{e_i^k} \qquad\qquad [2.7]
$$

where e_i^k are the exponents to which it is necessary to individually raise each fundamental dimension d_k so that the physical variable V_i is homogeneous with a product of fundamental dimensions raised to the exponents in question.

These exponents are shown in the dimensional matrix $\mathbf{D}$ presented in Figure 2.1.

	V_1	V_2	V_3	V_4	V_5	V_6
d_1	0	0	1	1	1	2
d_2	-2	-1	-3	0	-1	1
d_3	1	0	2	-3	-1	0

Figure 2.1. *Example 2.5: dimensional matrix D (black outline): the columns correspond to the physical variables V_i (light gray boxes) and the rows of fundamental dimensions d_k (dark gray boxes)*

In this example, (d_1, d_2 and d_3) is the initial base, made up of the fundamental dimensions, in which the set of physical quantities V_i is expressed.

2.1.3.3. *Step 3: applying the Vaschy–Buckingham theorem*

When the dimensional matrix has been written, the number of internal measures (or dimensionless numbers, denoted as π_i) which describe the cause–effect relationship of the physical process being studied can be determined. This can be done by applying the *Vaschy–Buckingham theorem*.

THEOREM 2.1.– Vaschy–Buckingham theorem

Any physical quantity representing a phenomenon (a target variable V_1) function of m *independent* physical quantities, V_i, measured by n_d fundamental dimensions[5], d_k, can be described by an implicit function between m-n_d dimensionless numbers π_i.

$$V_1 = f(V_2, V_3, V_4, ..., V_m) \text{ becomes } \pi_1 = F(\pi_2, \pi_3, \pi_4, ...,\pi_{m\text{-}nd})$$

with $\pi_i = \dfrac{V_i}{\Pi_j V_j^{a_{ij}}}$ [2.8]

where the exponents a_{ij} are rational numbers which may be zero, $i \in [1; m\text{-}n_d]$ and $j \in [m\text{-}n_d+1; m]$.

In this book, any implicit function F between the target dimensionless number and the other dimensionless numbers will be designated by the term *process relationship*.

Depending on the value of exponents a_{ij}, dimensionless numbers π_i are monomial groupings containing a minimum of one and up to n_d+1 physical quantities V_i.

As described in Chapter 1, the perspective offered by the Vaschy–Buckingham theorem is particularly interesting, namely when an experimental program is implemented to determine a process relationship F (equation [2.8]). Indeed, this theorem indirectly reveals that the number of experiments can be significantly reduced: a phenomenon described by a collection of m physical quantities can be described with this theorem as a collection of m-n_d internal measures. This is why this method is occasionally known as the "*method of reduced variables*".

---

5 These fundamental dimensions must come from a coherent system of units.

2.1.3.4. *Step 4: constructing the dimensionless numbers*

The set of dimensionless numbers π_i associated with the physical phenomenon being studied is generated using some transformations carried out on the dimensional matrix. To do so, it is necessary to highlight certain physical variables, which we will call *repeated variables*. Each dimensionless number making up the final set will be associated with a remaining physical variable, which we will call a *non-repeated variable*.

The set (or space) of dimensionless numbers, $\{\pi_i\}$, is obtained by dividing each non-repeated physical variable by a product of repeated variables raised to various exponents a_{ij} (these exponents may sometimes be equal to 0 depending on the dimension of the non-repeated physical variable).

$$\pi_i = \frac{V_{i,non-repeated}}{\prod_{j=1}^{n_d}\left[V_{j,repeated}\right]^{a_{ij}}} \quad \text{with } i \in [1; \text{m-n}_d] \tag{2.9}$$

Dimensionless numbers, therefore, represent *internal measures of non-repeated variables on the base made up of repeated physical variables*.

The shift from physical quantities to a space of dimensionless numbers which characterize the process is, therefore, carried out due to *a change of base*:

– the *initial base* is made up of the set of fundamental dimensions necessary for expressing the dimensions of the physical quantities listed;

– the *new base* is made up of repeated physical variables.

The choice of repeated variables is free, provided that:

– they are *dimensionally independent*;

– their dimensions *cover all the fundamental dimensions necessary for expressing the dimensions of the physical quantities listed*.

The notion of *dimensional independence* means that the dimension of a repeated physical variable cannot be expressed as a product of powers of the dimensions of the other repeated variables.

If, and only if, these two conditions are satisfied, then the number of repeated physical variables is equal to the number of fundamental dimensions necessary to express the physical quantities acting on the physical phenomenon. If this is not the case, a new set of repeated variables must be chosen in order to satisfy both conditions.

It is important to note that *the set of repeated variables satisfying these two conditions is not necessarily unique*, which may lead to several spaces of internal measures to describe the same phenomenon. This fact is not sufficiently underlined in the literature and is a cause of confusion. Indeed, a process engineer wishing to model his/her process through dimensional analysis commonly expects to obtain a single space of dimensionless numbers for a given problem and feels destabilized when he/she realizes that an infinity of spaces of internal measures is possible. This is the root cause of an unjustified loss of credit for dimensional analysis, where in the past, the construction of dimensionless numbers was not thought to be based on any theoretical framework even though the existence of these multiple spaces is mathematically proven (in work by Saint-Guilhem [SAI 62]).

The creation of a set of dimensionless numbers, therefore, relies on a particularly codified mathematical operation: the *change of base*. This consists of resolving a system of linear equations (whose unknowns are the exponents a_{ij} defined in equation [2.9]). This is a problem which is well known to mathematicians. Several methods are possible:

– the elimination/substitution method;

– the method based on the translation of a system of linear equations into matrix form, which we will call "resolution by matrix operations".

The latter method, although more abstract, should take priority when the system introduces many physical quantities in order to avoid cumbersome calculations (and associated errors!). Once the base of repeated variables is defined, the *choice of the resolution method has no influence on the dimensionless numbers formed*.

In the following sections, we provide the rules to adopt for each resolution method by using example 2.5. In the remainder of the book, these two methods will be used in the examples. This will also allow the user to note that these mathematical operations (be algebraic or matricial) are not insurmountable obstacles.

2.1.3.4.1. The elimination/substitution method

To explain the first method, we will start off with example 2.5.

EXAMPLE 2.5.– Fundamental rules for constructing the set of dimensionless numbers associated with a physical phenomenon: step 4 – elimination/substitution method (II)

The dimensional matrix associated with this example is presented in Figure 2.1 (the target variable is V_1).

First, it is necessary to choose the base constituted of the repeated variables whose number is equal to the number of fundamental dimensions necessary to express the physical phenomenon, which is 3 in this example. We then arbitrarily choose the following set of repeated variables: V_3, V_4 and V_5 (which we will call "base 1"). They are dimensionally independent and cover all the fundamental dimensions d_k necessary for expressing the dimensions of the physical quantities listed (Figure 2.1). Thus, we find recurrence of the variables V_3, V_4 and V_5 as the denominator of each dimensionless number formed, raised to different exponents. They constitute the new base, in which the dimensions of the non-repeated physical quantities will be expressed.

According to the Vaschy–Buckingham theorem, six quantities (V_i) measured by three fundamental dimensions (d_k) result in a relationship between $6 - 3 = 3$ dimensionless numbers. Therefore, the three numbers to determine, π_1, π_2 and π_3, which are, respectively, the internal measures of V_1, V_2 and V_6, are expressed using equation [2.9] as follows:

$$\pi_1 = \frac{V_1}{(V_3)^{a_{11}} \cdot (V_4)^{a_{12}} \cdot (V_5)^{a_{13}}} \qquad [2.10]$$

$$\pi_2 = \frac{V_2}{(V_3)^{a_{21}} \cdot (V_4)^{a_{22}} \cdot (V_5)^{a_{23}}} \qquad [2.11]$$

$$\pi_3 = \frac{V_6}{(V_3)^{a_{31}} \cdot (V_4)^{a_{32}} \cdot (V_5)^{a_{33}}} \qquad [2.12]$$

Since the π_i numbers are by definition dimensionless, it is thus sufficient to calculate the values of exponents a_{1j}, a_{2j} and a_{3j} with $j = 1, 2, 3$, by writing the following systems of three equations with three unknowns using the dimensional matrix (Figure 2.1):

– for π_1 :

$$\begin{cases} d_1: & 0 = a_{11} + a_{12} + a_{13} \\ d_2: & -2 = -3a_{11} + 0a_{12} - a_{13} \\ d_3: & 1 = 2a_{11} - 3a_{12} - a_{13} \end{cases}$$ [2.13]

– for π_2 :

$$\begin{cases} d_1: & 0 = a_{21} + a_{22} + a_{23} \\ d_2: & -1 = -3a_{21} + 0a_{22} - a_{23} \\ d_3: & 0 = 2a_{21} - 3a_{22} - 1a_{23} \end{cases}$$ [2.14]

– for π_3 :

$$\begin{cases} d_1: & 2 = a_{31} + a_{32} + a_{33} \\ d_2: & 1 = -3a_{31} + 0a_{32} - a_{33} \\ d_3: & 0 = 2a_{31} - 3a_{32} - a_{33} \end{cases}$$ [2.15]

The resolution of these systems results in:

$$\begin{cases} a_{11} = 3; & a_{12} = 4; & a_{13} = -7 \\ a_{21} = 2; & a_{22} = 3; & a_{23} = -5 \\ a_{31} = -8; & a_{32} = -13; & a_{33} = 23 \end{cases}$$ [2.16]

which then gives:

$$\left\{ \pi_{i,base\ 1} \right\} = \left\{ \pi_1 = \frac{V_1}{V_3^3 . V_4^4 . V_5^{-7}} , \pi_2 = \frac{V_2}{V_3^2 . V_4^3 . V_5^{-5}} , \pi_3 = \frac{V_6}{V_3^{-8} . V_4^{-13} . V_5^{23}} \right\}$$ [2.17]

Dimensional analysis, therefore, indicates that there is a *process relationship* between the dimensionless number containing the target variable ($\pi_1 = \pi_{target}$) and the dimensionless numbers π_2 and π_3, such that:

$$\frac{V_1}{V_3^3 . V_4^4 . V_5^{-7}} = F \left(\frac{V_2}{V_3^2 . V_4^3 . V_5^{-5}} , \frac{V_6}{V_3^{-8} . V_4^{-13} . V_5^{23}} \right)$$ [2.18]

2.1.3.4.2. Resolution method by matrix operations

The resolution method by matrix operations consists of three steps. After describing them, we will illustrate how they are implemented using example 2.5:

1) Defining core and residual matrices

Once the dimensional matrix D is established, the *core matrix*, denoted by C, must be identified. This submatrix of matrix D only corresponds to the repeated variables. It is, therefore, a square matrix whose size is determined by the number of fundamental dimensions necessary to describe the dimensions of the physical quantities listed.

By convention in this book, the core matrix C will be positioned in the final columns of the dimensional matrix D. There is no reason why the dimensional matrix would immediately contain the repeated variables positioned in the final columns. Therefore, it is necessary to reorganize the dimensional matrix in order to satisfy this convention.

The existence of the core matrix C, therefore, assumes that:

– a choice of repeated variables has already been made;

– the repeated variables chosen are dimensionally independent and that their dimensions cover all the fundamental dimensions necessary to express the dimensions of the physical quantities listed. *In order to verify this, we can calculate the determinant of the core matrix and ensure that it is not zero.*

The matrix adjacent to the core matrix C in the dimensional matrix D is known as the *residual matrix* and is denoted by R. Its columns are made up of physical quantities not chosen as repeated variables (that is to say the non-repeated variables). Unlike the core matrix, the residual matrix is not necessarily a square matrix.

2) Transformation of the core matrix into an identity matrix

Once the repeated variables have been chosen, and the dimensional matrix reorganized, the core matrix C must be transformed into an identity matrix (which we will hereby denote as C_I). During this transformation, the *residual matrix and, by extension, the dimensional matrix are modified*: we will denote, respectively, by R_m and D_m the residual and dimensional matrices obtained following the transformation of the core matrix into an identity matrix.

The matrix R_m will be useful for us to express the internal measures since it contains the required exponents a_{ij}. It is obtained from operations which use the core matrix C and the residual matrix R. Indeed, the m-n_d systems, each made up of n_d equations for n_d unknowns, written during the implementation of the elimination/substitution method can be put into the following matrix form:

$$R = C \times R_m \qquad\qquad [2.19]$$

where:

– " $\times$ " corresponds to the product of the two matrices;

– R_m is the matrix which contains the required exponents a_{ij}. This is, therefore, nothing else but the modified residual matrix.

In order to determine the exponents a_{ij} contained in R_m, it is thus necessary to transform equation [2.19] as follows:

$$R_m = C^{-1} \times R \qquad\qquad [2.20]$$

where C^{-1} is the inverse matrix of C. This transformation is equivalent to applying the Gauss pivot method.

It should be noted that it is possible to transform the core matrix into an identity matrix by another method, which consists of carrying out linear combinations of each row.

3) Determining dimensionless numbers using the modified residual matrix

Once the core matrix is transformed into an identity matrix, the dimensionless numbers characterizing the physical phenomenon can be written using the coefficients contained in the columns of the modified residual matrix R_m. Each non-repeated variable (i.e. each column of the modified residual matrix) forms a dimensionless number π_i. *Each dimensionless number contains in the numerator a non-repeated variable and in the denominator the chosen repeated variables raised to the powers indicated in the modified residual matrix R_m.*

EXAMPLE 2.5.– Fundamental rules for constructing a set of dimensionless numbers associated with a physical phenomenon: step 4 – resolution method by matrix operations (III)

The dimensional matrix associated with this example is presented in Figure 2.1. The target variable is V_1 and the repeated variables are V_3, V_4 and V_5.

The systems of three equations containing three unknowns described in equations [2.13]–[2.15] can be put into a matrix form as follows:

$$\underbrace{\begin{bmatrix} 0 & 0 & 2 \\ -2 & -1 & 1 \\ 1 & 0 & 0 \end{bmatrix}}_{\mathbf{R}} = \underbrace{\begin{bmatrix} 1 & 1 & 1 \\ -3 & 0 & -1 \\ 2 & -3 & -1 \end{bmatrix}}_{\mathbf{C}} \times \underbrace{\begin{bmatrix} a_{11} & a_{21} & a_{31} \\ a_{12} & a_{22} & a_{32} \\ a_{13} & a_{23} & a_{33} \end{bmatrix}}_{\mathbf{R_m}}$$

[2.21]

The matrix of the first member of equation [2.21] corresponds to the non-repeated variables (residual matrix $\mathbf{R}$), whereas the second member includes, first, the matrix of the repeated variables (core matrix $\mathbf{C}$) and, second, the matrix of the exponents a_{ij} required (modified residual matrix $\mathbf{R_m}$ required).

It is possible to identify the core and residual matrices using a reorganized dimensional matrix (Figure 2.2). This latter matrix is deduced from the dimensional matrix described in Figure 2.1. We have simply reorganized it in order to position the three repeated variables in the final columns.

	V_1	V_2	V_6	V_3	V_4	V_5
d_1	0	0	2	1	1	1
d_2	-2	-1	1	-3	0	-1
d_3	1	0	0	2	-3	-1

Figure 2.2. *Example 2.5: identification of repeated variables forming the core matrix $\mathbf{C}$ (dark gray boxes) and the residual matrix $\mathbf{R}$ (light gray boxes) within the dimensional matrix $\mathbf{D}$ (black outline)*

In Figure 2.2, the core matrix $\mathbf{C}$ is the submatrix 3×3 shown in dark gray, and the residual matrix $\mathbf{R}$ is the submatrix shown in light gray[6].

After the transformation of the matrix as described in equation [2.20], the modified dimensional matrix $\mathbf{D_m}$ is obtained. It is shown in Figure 2.3.

	V_1	V_2	V_6	V_3	V_4	V_5
$-2d_2-3d_1-d_3$	3	2	-8	1	0	0
$-5d_1-2d_3-3d_2$	4	3	-13	0	1	0
$9d_1+3d_3+5d_2$	-7	-5	23	0	0	1

Figure 2.3. *Example 2.5: production, using equation [2.20] or linear combinations of the new dimensional matrix $\mathbf{D_m}$ (black outline), the identity matrix $\mathbf{C_I}$ (dark grey boxes) and the modified residual matrix $\mathbf{R_m}$ (light grey boxes)*

It would also have been possible to transform the core matrix into an identity matrix using linear combinations between the rows of the dimensional matrix. For information, we have listed these combinations in Figure 2.3 in the first column of the modified dimensional matrix $\mathbf{D_m}$.

By using the coefficients contained in the modified residual matrix $\mathbf{R_m}$ (Figure 2.3), the following three dimensionless numbers are deduced:

$$\left\{ \pi_{i,base\ 1} \right\} = \left\{ \pi_1 = \frac{V_1}{V_3^3 . V_4^4 . V_5^{-7}} , \pi_2 = \frac{V_2}{V_3^2 . V_4^3 . V_5^{-5}} , \pi_3 = \frac{V_6}{V_3^{-8} . V_4^{-13} . V_5^{23}} \right\} \qquad [2.22]$$

The dimensionless numbers which appear in equation [2.22] are indeed identical to those obtained with the elimination/substitution method (equation [2.17]).

2.1.3.4.3. Shift relationships between dimensionless number spaces depending on sets of chosen repeated variables

As previously stated, the set of repeated variables must satisfy two conditions (dimensional independence and coverage of all the fundamental dimensions necessary for expressing the dimensions of the physical

---

6 We will use this color coding system throughout the book.

quantities listed), but it is not unique. Ultimately, several sets of dimensionless numbers (and process relationships characterizing the physical phenomenon), valid from a dimensional analysis point-of-view, may be obtained, without there being one which has priority over the others. *This multitude of spaces of dimensionless numbers is not problematic given that there are shift relationships between these spaces.* Szirtes [SZI 07] showed that a set of dimensionless numbers obtained from a base can be expressed as combinations of dimensionless numbers obtained from another base by introducing a shift matrix **P**:

$$
\begin{bmatrix}
\ln\left(\pi_{1,base2}\right) \\
\ln\left(\pi_{2,base2}\right) \\
\vdots \\
\ln\left(\pi_{(m-n_d),base2}\right)
\end{bmatrix}
= \mathbf{P} \times
\begin{bmatrix}
\ln\left(\pi_{1,base1}\right) \\
\ln\left(\pi_{2,base1}\right) \\
\vdots \\
\ln\left(\pi_{(m-n_d),base1}\right)
\end{bmatrix}
\tag{2.23}
$$

where $\{\pi_{i,\,base\,1}\}$ and $\{\pi_{i,\,base\,2}\}$ designate the sets of dimensionless numbers generated from the dimensional matrix **D** using bases 1 and 2, respectively. According to the convention used in this book (repeated variables being placed in the final columns of the dimensional matrix), $\pi_{k,\,base\,2}$ therefore designates the dimensionless number π_i constructed using the kth column of the dimensional matrix obtained with base 2.

The shift matrix **P** which appears in equation [2.23] is defined by:

$$
\mathbf{P} = (\mathbf{S}_1 - \mathbf{R}_{m,base1}{}^T \times \mathbf{S}_3)^{-1}
\tag{2.24}
$$

where " -1" is the inverse matrix, "T" is the transpose of the matrix, $\mathbf{R}_{m,\,base\,1}$ is the modified residual matrix obtained with base 1 and $\mathbf{S}_1$ and $\mathbf{S}_3$ are the submatrices contained in matrix **A**. These final three matrices are constructed in the following way:

– the rows of matrix **A** reproduce the order of appearance of the physical quantities in the columns of the dimensional matrix associated with base 1;

– the columns of matrix **A** reproduce the order of appearance of the physical quantities in the columns of the dimensional matrix associated with base 2;

– the coefficients of matrix **A** consist solely of 0's and 1's. Number 1 appears when the same physical quantity appears at the top of the row and column;

– the rows of the submatrix S_1 reproduce the order of appearance of the non-repeated physical variables on base 1. The columns of the submatrix S_1 reproduce the order of appearance of the non-repeated physical variables on base 2. The submatrix S_1 is, therefore, located at the top left of matrix **A**;

– the rows of the submatrix S_3 reproduce the order of appearance of the repeated physical variables on base 1. The columns of the submatrix S_3 reproduce the order of appearance of the non-repeated physical variables on base 2. The submatrix S_3 is, therefore, located under the submatrix S_1, thus at the bottom left of matrix **A**.

An illustration of these shift relationships is shown in Appendix 1, using example 2.5 again.

2.1.3.4.4. Conclusion

It should be noted that:

– once the base of the repeated variables is chosen, there are various methods for carrying out the change of base necessary for determining the set of dimensionless numbers;

– there is not a single set of dimensionless numbers which characterize the physical phenomenon, but several sets depending on the repeated variables chosen to make up the base;

– it is nevertheless possible to move from one set to another through matrix operations.

Finally, the readers must remember that dimensional analysis helps to construct the set of dimensionless numbers associated with a physical phenomenon, but it in no way indicates the mathematical form of the process relationship F linking the dimensionless number containing the target variable (π_{target}) and the other dimensionless numbers π_j. The latter can be evaluated, with greater or lesser accuracy, by a more or less substantial experimental program. It will then be necessary to make tests by varying the dimensionless numbers π_j and to measure the consequences of these variations on π_{target}. We will refer back to this with the help of various resolved examples (see Chapter 6).

2.1.3.5. *Step 5: rearranging dimensionless numbers*

We have seen that various spaces of dimensionless numbers characteristic of a phenomenon coexist, depending on the base chosen. It is possible to modify the space of dimensionless numbers obtained by consistent rearrangements which consist of raising certain dimensionless numbers to different powers, multiplying them with each other, adding and/or subtracting them. Indeed, a dimensionless number raised to a power, a product, a quotient, a sum or a difference of dimensionless numbers still remains a dimensionless number.

As a result, it is always possible to substitute an internal measure π_i of a space of dimensionless numbers by:

– this same internal measure raised to any power $(\pi_i)^\alpha$;

– the product of this internal measure with other dimensionless numbers, whether or not it is raised to a power $(\pi_j)^\alpha \times (\pi_i)^\beta$;

– the sum of this internal measure with a dimensionless constant (denoted as a) or with other dimensionless numbers, whether or not they are raised to a

power $a - \dfrac{\left(\pi_i\right)^\alpha}{\left(\pi_j\right)^\beta}$ or $(\pi_j)^\alpha + (\pi_i)^{\beta^\sim}$

These rearrangements of internal measures, occasionally known as *recombinations*, are generally carried out to:

– *give rise to certain common dimensionless numbers* whose physical meaning is well established (ratio of the fluxes of mass, momentum, energy, retentions, etc.). Indeed, as mentioned in Chapter 1, one of the major reasons for carrying out dimensional analysis is to gain a phenomenological understanding of the process being studied;

– *eliminate fractional exponents*;

– eliminate a physical quantity of a dimensionless number in order to obtain an internal measure independent of it, or to isolate a physical quantity (particularly the target variable) within a single internal measure. This allows the researcher to evaluate with greater ease how the variations in this physical quantity directly influence the target variable.

It is important to note that the rearrangement of the dimensionless numbers does not entail a reduction in the number of internal measures which characterize the physical phenomenon.

The interest of these rearrangements of dimensionless numbers will be illustrated through examples given in Chapter 6.

2.1.3.6. Conclusion

We can note that *the fundamental rules for constructing a set of dimensionless numbers which have been established over five steps require no conditions other than the coherence of the system of units.* This means that this process is not limited to the SI (even though it is the system we recommend!), but can be applied with any coherent system of units.

2.2. From internal measures to the form of a process relationship

2.2.1. Choosing a set of internal measures suitable for the experimental program used

In sections 2.1.3.4.3 and 2.1.3.5, we saw that for a given problem defined by a set of physical quantities:

– several sets of dimensionless numbers can be immediately formed depending on the choice of repeated variables;

– it is also possible to rearrange the dimensionless numbers obtained at will.

Therefore, there is an infinite number of complete sets of dimensionless numbers. The underlying question is to know whether the number of dimensionless numbers π_i relating to causes may be minimized before the experimental program needed to establish the process relationship F of π_{target} takes place, with a view to minimizing the number of experiments.

Using mathematical considerations (invariance of vector relationships in some groups of affinities), Saint-Guilhem [SAI 62] demonstrated that the Vaschy–Buckingham theorem mentioned earlier does not necessarily lead to a minimization of dimensionless numbers π_i defining the problem. He showed and tracked the method to achieve this. This consists of choosing a system of units which is adapted to the problem being studied. However, making appropriate use of this method is not easily accessible to non-experts. This is why, in this book, we will not discuss this approach to ensure that the dimensionless numbers formed constitute the minimal set. As an initial approach, we will not attempt to use any system other than SI, and consequently accept dealing with a number of internal measures which are not necessarily minimal. Nevertheless, Chapter 3 highlights certain techniques which can be used to reduce the number of internal measures and get closer to this minimal number.

The next question is to know whether certain sets $\{\pi_i\}$ can be used more practically than others in order to understand the nature of the process relationship linking π_{target} to the other terms π_i.

As mentioned by Buckingham [BUC 21], the physical analysis is facilitated when *each physical quantity which can be experimentally controlled only appears in a single internal measure π_i.* Indeed, if all the dimensionless numbers π_i vary at the same time, the user may struggle to identify and understand the influence of a physical quantity on the target variable, independently from the other quantities.

In a practical sense, this means that it is desirable:

– to avoid (as much as possible) choosing as repeated variables those physical quantities which may have a significant influence on the physical phenomenon;

– to ensure that the target variable only appears in a single dimensionless number (and to therefore prevent it from appearing in the core matrix);

– to carry out rearrangement of dimensionless numbers in order to eliminate a physical quantity of certain internal measures, as mentioned in section 2.1.3.5.

It is clear that a compromise must be found between these recommendations and the interest of considering the usual dimensionless numbers (e.g. Reynolds number, Froude number, etc.) for which there are points of reference and a well-established physical meaning. We will return to this during the presentation of the examples.

2.2.2. What form of mathematical relationship should be chosen to correlate the dimensionless numbers?

2.2.2.1. General case: there is no information on the form of the process relationship

Dimensional analysis can give rise to internal measures which may influence the process being studied (the causes). However, dimensional analysis:

– does not show in what proportions and what ranges of operating conditions the internal measures influence the process being studied. Indeed, some internal measures can have a very significant effect, while others have a very small effect;

– nor does it give the mathematical form of the process relationship correlating the dimensionless numbers relating to the causes with the target internal measure. The only exception is when a single dimensionless number defines the problem (this case is dealt with separately in section 2.2.2.2).

Only an idea regarding the underlying mechanisms (analytical or numerical knowledge of the solution) and/or the experiment can help to approximate the most appropriate mathematical expression for the process relationship. We will see that due to a lack of information on these elements, it is generally considerations regarding the facilities of mathematical adjustment which cause one type of function to be preferred over another.

It can also be useful to use the (analytical or empirical) mathematical models available in the literature in order to identify the form of the process relationship which can describe our own experimental data. It is, therefore, essential to examine *these models' range of validity*, namely:

– is the list of physical quantities established in these models relevant with regard to the problem being studied?

– can the physical problem in question be defined by the same number of internal measures?

– for which intervals of internal measures have they been validated?

If the model's range of validity covers that of the case being studied, it is then likely (depending on the errors of measure of the target variable) that the form of the process relationship is appropriate. Nothing is guaranteed in the other cases, even if only one of the values of the internal measures defining the causes of the phenomenon being studied differs from the range explored in the model. It is, however, possible to verify whether the model taken from the literature can be extrapolated to the conditions of the study or whether its mathematical form remains valid but needs to be adjusted to new parameters.

In the absence of known models, the search for the mathematical form of the process relationship is more delicate.

Consider the process relationship F linking the target internal measure, $\pi_l = \pi_{target}$, to the set of internal measures which define the causes of the phenomenon, giving:

$$\pi_{target} = F\left(\pi_2, \pi_3, \pi_4, ..., \pi_{m-n_d}\right) \qquad [2.25]$$

Initially, a relatively general form needs to be chosen. This is a form based on a *sum of monomials* where each *monomial* is a product of the internal measures raised to different powers:

$$\pi_{target} = \sum_1^p A_p.(\pi_2^{b_{2,p}}.\pi_3^{b_{3,p}}.\pi_4^{b_{4,p}}...\pi_{m-n_d}^{b_{(m-n_d).p}}) \qquad [2.26]$$

where m is the number of physical quantities V_i and n_d is the number of fundamental dimensions. The unknowns in equation [2.26] are exponents $b_{j,p}$ and constants A_p defined for each value of p (p integer). The higher the value of p, the more faithfully function F will describe the experimental points. The task of identification is complex and requires a great number of experimental data as $(m-n_p).p$ unknowns need to be determined.

In order to simplify this task and minimize the number of unknowns to be identified, the following relationship containing *a single monomial* is often preferred:

$$\pi_{target} = A_1.\pi_2^{b_2}.\pi_3^{b_3}.\pi_4^{b_4}...\pi_{m-n_d}^{b_{m-n_d}} \qquad [2.27]$$

In this case, only $m-n_d$ unknowns need to be determined by the experiments[7]. We will refer to this type of mathematical expression as a *monomial form*.

It is important to highlight the fact that nothing guarantees that the process relationship can be written in a monomial form (equation [2.27]), or is able to adjust the "true" physical law (which is theoretical but analytically inaccessible). It is, therefore, by default, and also because of their ability to approximate various families of mathematical functions, that these monomial forms are generally encountered in correlations.

---

7 These unknowns (A_1 and b_j) are obtained by minimizing a criterion generally based on the sum of the squares of the differences between the experimental values of π_{target} and those predicted by equation [2.27].

It is important at this juncture to reiterate a rule of correct usage in search of process relationship with a monomial form. It is essential to use an experimental program where each of the dimensionless numbers π_i varies over a sufficiently wide range. If this is not the case, the value of the associated exponents is not guaranteed and can introduce bias into the value of the constant A_1. The robustness of the process relationship, when the established range of validity is exceeded, will be even more limited.

This observation is illustrated using various examples presented in Chapter 6.

2.2.2.2. Specific case: a set of dimensionless numbers reduced to a singleton

Dimensionless analysis indicates that there is a process relationship F between the target dimensionless number ($\pi_1 = \pi_{target}$) and the other internal measures π_j related to the causes:

$$\pi_{target} = F\left(\pi_2, \pi_3, \pi_4, ..., \pi_{m-n_d}\right) \qquad [2.28]$$

Equation [2.28] can also be written in the following form:

$$\Psi\left(\pi_{target}, \pi_2, \pi_3, \pi_4, ..., \pi_{m-n_d}\right) = \text{const} \qquad [2.29]$$

In the specific case, where $m - n_d = 1$, equation [2.29] becomes:

$$\Psi\left(\pi_{target}\right) = \text{const} \qquad [2.30]$$

The space of dimensionless numbers is thus reduced to a single internal measure (π_{target}), known as a *singleton*. Equation [2.30] necessarily implies that this *singleton is a dimensionless constant*:

$$\pi_{target} = \text{const} \qquad [2.31]$$

The specific case of singletons will be illustrated in Chapter 3.

2.3. Configuration and operating point of a system

In the previous sections, a number of definitions, theorems and rules were set out in order to obtain the set of the dimensionless numbers relating to a physical phenomenon. This quasi-mechanical procedure is often known as *"blind" dimensional analysis*, even though this term is not completely suitable. Indeed, establishing the list of the relevant independent physical quantities which influence the target variable cannot be done without a minimum knowledge of the physical phenomena present (we will refer back to this point in Chapter 3 and throughout the examples).

The aim of this section is to:

– show how the dimensionless numbers obtained by "blind" dimensional analysis represent physical indicators of the system;

– propose a method of representation.

2.3.1. *Definition of the configuration of the system and its links with internal measures*

Introducing the notion of internal measures allowed us to show that the dimensionless numbers obtained through "blind" dimensional analysis correspond to the measures of a specific physical quantity, using as "standards" the physical quantities of the system itself (repeated variables). Consequently, it is then a question of generalizing this notion, especially by highlighting that these dimensionless numbers correspond to general features (ratios of fluxes and retentions, and geometric ratios) which can be found in a generic way in any system studied.

To do this, it is necessary to define the *configuration of the system*. This notion was introduced by Johnstone and Thring [JOH 57], and then was greatly developed upon by Becker [BEC 76]. It was more recently taken up by Midoux [MID 81]. We will examine this notion here.

The *configuration of the system*, made up of a control volume which is itself limited by a control surface, can be examined using one of the two approaches:

– either using potential fields;

– or using interactions with its environment.

The first approach consists of defining a system using potential fields (intensive quantities) which characterize it:

– geometrically, by boundary surfaces, moving over time or not;

– kinematically, by field of velocity $\vec{U}(\vec{X},t)$;

– thermally, by field of temperature $T(\vec{X},t)$;

– chemically, by field of compositions $x_i(\vec{X},t)$.

The drawback of this type of representation is that it masks the interactions of the system with its environment, while the distributions of the intensive quantities and their spatiotemporal evolutions are directly linked to the exchanges with this environment. For this reason, it is preferable to define the configuration of the system by explaining the interactions of the system with its environment (Figure 2.4), which is to say:

– in geometric terms by *boundary surfaces* moving over time or not, $S(t)$;

– in terms of *fluxes* of mass, momentum and energy;

– in terms of retentions by the extensive quantities within the control volume.

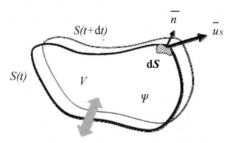

Flux exchange with environment

Figure 2.4. *Definition of a system made up of a control volume V, delimited by a boundary surface S in interaction with its environment*

As for any physical quantity, these fluxes, retentions and surfaces of the control volume, which define the configuration of the system, can be made dimensionless. This requires conservation equations to be written, or what is commonly known as balances, which is to say the quantification of inputs and outputs, and retentions of the system in terms of extensive variables (mass energy, momentum). Using as an internal standard the repeated physical variables which belong to the system, it is possible to show [BEC 76, MID 81] that the obtained dimensionless numbers are internal measures of fluxes and retentions, and geometric ratios. The base elements of this approach, known as *configurational analysis* or the *Pythagorean method*[8], are presented in Appendix 2. In this appendix, the dimensionless numbers commonly used in agro-food processes are expressed in the form of flux relationships (of mass, momentum and energy), retention relationships or geometric relationships. The expression of these relationships is directly deduced from conservation equations. One of the advantages of configurational analysis is to make explicit the physical meaning of these dimensionless numbers. Appendix 2 simply presents them to show this aspect. We will not use them as a complete method for constructing a space of dimensionless numbers characteristic of a system. According to Becker [BEC 76], this method, apart from the advantage previously mentioned, would help to overcome the inherent limitations of "blind" dimensional analysis.

The result of these considerations is that the internal measures, taken from a "blind" dimensional analysis of the system, are nothing other than those which would have appeared if we had undertaken a configurational analysis, which is to say:

1) define fluxes, retentions and geometry which characterize the system studied, using conservation equations;

2) make them dimensionless with an *ad hoc* base;

3) and finally, rearrange them if necessary.

This is why, in this book, we will call the *configuration of the system* a complete set (among those which are possible) of internal measures, $\{\pi_j$ with $\pi_j \neq \pi_{target}\}$, characterizing the *causes* responsible for the variation of the target internal measure.

---

8 In a rigorous process, the Pythagorean method is an integral part of the configurational analysis, which includes the theory of groups. We will not examine this aspect in detail in this book, and we refer to Becker [BEC 76].

To this mathematical relationship defining the configuration of a system, we will also associate a set of numerical values, which will be called *operating points*. An operating point is, therefore, a specific set of numerical values of internal measures $\{\pi_j$ with $\pi_j \neq \pi_{target}\}$ defining the causes responsible for the variation of the system.

This notion of the operating point is essential since to *each operating point is associated a physical variation of the system, and therefore an evolution of the target variable.*

We will illustrate these notions by returning yet again to example 2.5.

EXAMPLE 2.5.– Fundamental rules for constructing the set of dimensionless associated with a physical phenomenon: configuration and operating points (IV)

According to equation [2.22], it can be stated that the configuration of the system contains the following two internal measures which define the causes:

$$\pi_2 = \frac{V_2}{V_3^2 . V_4^3 . V_5^{-5}} \text{ and } \pi_3 = \frac{V_6}{V_3^{-8} . V_4^{-13} . V_5^{23}}$$

These two measures are responsible for the variation in the measures of the target internal measure π_1. If $V_2 = 2$, $V_6 = 3$ and $V_3 = V_4 = V_5 = 1$, then $\{\pi_2 = 2$, $\pi_3 = 3\}$ is a first operating point; if $V_2 = V_3 = V_4 = V_5 = V_6 = 1$, then $\{\pi_2 = 1$, $\pi_3 = 1\}$ is another operating point.

2.3.2. *Graphic representation of the operating points of a system*

Methods of graphic representation are proposed in order to provide a concise view of the operating points of a system. This type of representation will be a visual tool in order to:

– compare the various operating points of a system which have led to the different evolutions of the target variable;

– visualize the level of each internal measure characterizing the system, and thus understand its specific influence on the target variable.

Imagine that both internal measures π_2 and π_3 characterize the configuration of the system. If they vary during the experimental program in the intervals of the respective measures $\pi_2 \in [\pi_2 \text{ min}, \pi_2 \text{ max}]$ and $\pi_3 \in [\pi_3 \text{ min}, \pi_3 \text{ max}]$, then the set of operating points achievable by the system is grouped together inside a surface plane (Figure 2.5).

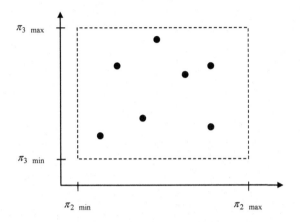

Figure 2.5. *Graphic representation of the configuration of the system and operating points explored when two internal measures characterize the configuration of the system*

If the number of internal measures representing the causes of the phenomenon is higher than three, the graphic representation is not direct. Imagine, therefore, that six dimensionless numbers { π_2, π_3, π_4, π_5, π_6, π_7 } represent the configuration of the system. For a given set of operating conditions, these internal measures will take six values, and mathematically constitute a 6-tuple[9]. In Figure 2.6, a star-shaped representation has been adopted whereby each axis of the star is associated with an internal measure. The value of each internal measure is shown by a black spot, and the intervals in which each internal measure π_j varies during the experiment are marked by gray bars with limits at each end. It is possible to join up these black spots with segments in the ascending order of the numbers of the internal measures ($\pi_2 \rightarrow \pi_3 \rightarrow \pi_4 \rightarrow \pi_5 \rightarrow \pi_6 \rightarrow \pi_7 \rightarrow \pi_2$).

---

9 A 6-tuple is an ordered collection of six objects.

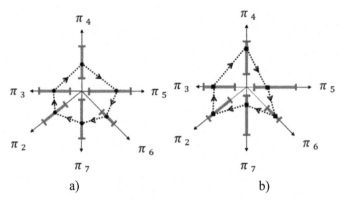

Figure 2.6. *Graphic representation of the configuration of the system and operating points explored when six internal measures characterize the configuration of the system*

By adopting these conventions, the operating point is represented in Figure 2.6 by a black polygon connecting the various "tuples" (black spots in Figure 2.6(a)). A further example of another operating point would be the black polygon linking the black squares in Figure 2.6(b).

So as to lighten the graphic representations of operating points explored, we decided in this book to leave only the *intervals of variations for each internal measure* (gray bars with limits at both ends) defining the configuration of the system.

2.3.3. Conclusion

The essential nature of these notions on the configuration of the system, operating points and their graphic representation will be fully highlighted:

– in Chapter 5, which deals with the issue of scale change. Above all, we see that the concepts of complete or partial similarity rely on the total or partial conservation of the operating point on both scales;

– with the help of various examples set out throughout the book.

At this stage, it is important to say that the configuration of the system (i.e. the number of internal measures responsible for the causes of the variation of the target internal measure) as set out in this chapter can be reduced to what we can call a *reduced configuration*. However, under such a term, very different situations can be grouped. Indeed:

– we will examine in Chapter 3 that there are different techniques which can be used to reduce the configuration of the system (case 1). These techniques are based on the introduction of an intermediate variable, on the splitting of a fundamental dimension into two fundamental dimensions of the same nature, known as "dimension splitting", or on the introduction of a supplementary fundamental dimension;

– certain *constraints* lead to a reduction in the configuration of the system as obtained by the Vaschy–Buckingham theorem. These constraints have different natures since they can be imposed:

- either by conditions directly imposed on certain physical quantities (case 2). For instance, the geometric quantities which characterize the boundaries of the system are fixed;

- or by experiments conducted to establish a process relationship which will not allow for variation of certain internal measures in sufficiently wide ranges (case 3). For instance, this may be the roughness of a pipe.

Nevertheless, it is important to note that *the consequences of these various reductions in the configuration of the system on the process relationship are not of the same order*. In case 1, the reduction does not lead to a restriction of the range of validity for the process relationship, while this does happen for cases 2 and 3.

We, therefore, invite the readers to remain vigilant with regard to this notion of reduced configuration. In this book, we will use it sparingly. In all the examples set out, we will specify exactly, if applicable, what is intended by the reduced configuration.

2.4. Guided example 1: power consumption for a Newtonian fluid in a mechanically stirred tank

We will address the following example so as to show how to practically implement the principles and rules stated in the previous sections, and to illustrate their application in a simple case of agro-food processes.

Here, we are examining the power demands of an engine block driving an impeller, vertically centered, in a tank with a flat bottom. To do this, it is necessary to determine the power required to ensure the rotation of the impeller according to the characteristics of the Newtonian fluids being used and the range of the impeller rotational speeds planned.

This problem has no theoretical solution if the geometry of the agitation system is not simple, that is to say, if the ideal system of vertically centered coaxial cylinders is not used. Of course, it is possible, using knowledge of fluid mechanics (conservation equations), to not omit any physical quantity influencing the process. In order to provide an illustrative example, here we purposefully attempt to tackle this problem using "blind" dimensional analysis, by finding the physical quantities which influence the power on the sole basis of the "common sense".

2.4.1. *List of relevant independent physical quantities*

The first step involves:

– defining the *target variable*. In this example, this is the power consumed by the impeller (denoted as P) which is supposed to correspond to the power provided by the engine block;

– listing the independent physical quantities which influence the target variable.

To help us, Figure 2.7 brings together the variables involved in the problem. A simplified geometry has been chosen here: the cylindrical tank has no baffles and is flat-bottomed (so only two lengths are necessary to define the geometry of the tank), and the impeller is vertically centered in the tank. Moreover, the agitation operation is considered as being isothermal.

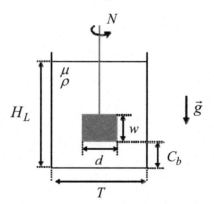

Figure 2.7. *Guided example 1: graphic representation and identification of physical quantities influencing the power consumed by the impeller in a mechanically stirred tank*

A simple method to help list the physical quantities which influence the process being studied consists of classifying them in terms of *material parameters, process parameters, boundary and initial conditions,* and *universal constants*[10]. This is the method which we will apply in this example.

According to the graphic representation in Figure 2.7, the *boundary conditions* which may have an influence on the power consumed by the impeller can be summarized with the following geometric parameters: T, d, C_b, H_L and w. Indeed, when one of these quantities varies, the friction within the flow domain is modified, and leads to an increase or reduction in power.

As the field of gravity acts on the free surface of the fluid, acceleration g should not be forgotten (although it is constant in all experiments).

The flow domain contains a homogeneous Newtonian fluid. The *material parameters* are its dynamic viscosity (μ) and its density (ρ).

Finally, the *process parameter* which has a decisive effect on the behavior of the system is the rotational speed of the impeller (N).

The physical phenomenon being studied is, therefore, summarized in the relationship:

$$P = \mathrm{f}\ (T, d, C_b, H_L, w, \mu, \rho, N, g) \tag{2.32}$$

The physical quantity of the first member (P) is the result of the nine physical quantities of the second member which group together all of the causes.

2.4.2. *Determining the dimensions of the physical quantities*

The second step involves determining the dimensions of the physical quantities previously listed. Beforehand, the number of fundamental dimensions and units must be established, or in other words, the unit system must be defined. For reasons mentioned earlier, the SI will be adopted. The dimension of each physical quantity listed can thus be expressed in terms of

---

10 We will refer back to this in more detail in Chapter 3.

fundamental dimensions. From this, the dimensional matrix can be constructed (Figure 2.8).

	P	μ	N	d	w	H_L	C_b	ρ	g	T
M	1	1	0	0	0	0	0	1	0	0
L	2	-1	0	1	1	1	1	-3	1	1
T	-3	-1	-1	0	0	0	0	0	-2	0

Figure 2.8. *Guided example 1: dimensional matrix **D** (black outline): the columns correspond to physical variables (light gray boxes) and the rows correspond to fundamental dimensions (dark gray boxes)*

2.4.3. Application of the Vaschy–Buckingham theorem and the construction of dimensionless numbers

Only three fundamental dimensions were necessary for expressing the dimensions of all the physical quantities listed. According to the Vaschy–Buckingham theorem, the final number of internal measures will be equal to seven (10 physical quantities listed, minus three fundamental dimensions).

Three physical quantities are, therefore, required to make up a base. We will choose the three quantities ρ, N and d as *repeated variables* as far as:

– they are *dimensionally independent*, where none of these three variables can be expressed as a product of powers of dimensions of the two other variables (the determinant of the associated matrix is not zero);

– they cover all the fundamental dimensions necessary to express the dimensions of the physical quantities listed.

This choice of the base (ρ, N and d) is arbitrary, and other choices would have been possible[11]. It should be remembered that this is not a problem because there are shift relationships between spaces of dimensionless numbers (see 2.3.4.3).

---

11 We will refer back to this guided example in Chapter 3 using bases other than (ρ, N and d).

The reorganization of the dimensional matrix **D** following the choice of these three independent repeated variables (ρ, N and d) is represented in Figure 2.9.

	P	μ	g	T	w	H_L	C_b	ρ	N	d
M	1	1	0	0	0	0	0	1	0	0
L	2	-1	1	1	1	1	1	-3	0	1
T	-3	-1	-2	0	0	0	0	0	-1	0

Figure 2.9. *Guided example 1: the identification of the repeated variables forming the core matrix* **C** *(dark gray boxes) in the dimensional matrix* **D** *(black outline), the residual matrix* **R** *(light gray boxes)*

As we have seen (see section 2.1.3.4), the creation of a set of dimensionless numbers consists of a change of base: here, changing from base (M, L and T) to base (ρ, N and d). To resolve this mathematical problem, various techniques are available to us. With regard to the number of physical quantities listed (10), the elimination-substitution method is not the most suitable, thus making the method using matrix operations more relevant. We can transform the core matrix into an identity matrix, either by applying equation [2.20] or with the technique of linear combinations between rows. In this case, we choose the second option. Indeed, by inverting rows L and T in matrix **D**, the combinations between the rows are very simple, as shown in Figure 2.10.

	P	μ	g	T	w	H_L	C_b	ρ	N	d
M	1	1	0	0	0	0	0	1	0	0
-T	3	1	2	0	0	0	0	0	1	0
L+3M	5	2	1	1	1	1	1	0	0	1

Figure 2.10. *Guided example 1: creating the modified dimensional matrix* **Dm** *(black outline) with the identity matrix* **C₁** *(dark gray boxes) and the modified residual matrix* **Rm** *(light gray boxes)*

The coefficients contained in the modified residual matrix **R**$_m$ obtained in Figure 2.10 then allow internal measures to be formed. Therefore, the dimensional analysis transforms equation [2.32] into:

$$\pi_{target} = \frac{P}{\rho.N^3.d^5}$$

$$= F\left(\pi_2 = \frac{\mu}{\rho.N.d^2}, \pi_3 = \frac{g}{N^2.d}, \pi_4 = \frac{T}{d}, \pi_5 = \frac{w}{d}, \pi_6 = \frac{H_L}{d}, \pi_7 = \frac{C_b}{d}\right) \qquad [2.33]$$

Six internal measures $\{\pi_2, \pi_3, \pi_4, \pi_5, \pi_6, \pi_7\}$ define the configuration of the system and appear as responsible for the variations of the target variable. The latter is commonly known as the power number (or Newton number) and denoted by N_P (or Ne).

2.4.4. Rearrangements of dimensionless numbers

We have already discussed (see section 2.1.3.5) how it is always possible to raise an internal measure to any power. Here, by raising π_2 and π_3 to the power -1, the Reynolds and Froude numbers for stirring appear as:

$$(\pi_2)^{-1} = Re = \frac{\rho.N.d^2}{\mu}, \quad (\pi_3)^{-1} = Fr = \frac{N^2.d}{g} \qquad [2.34]$$

These numbers have a known physical meaning, well suited to the flow of a homogeneous Newtonian fluid: they, respectively, represent the ratio of inertial and viscous forces, and inertial and gravitational forces (see Appendix 2). Equation [2.32] becomes:

$$\pi_{target} = \frac{P}{\rho.N^3.d^5} = F\left(Re, Fr, \pi_4 = \frac{T}{d}, \pi_5 = \frac{w}{d}, \pi_6 = \frac{H_L}{d}, \pi_7 = \frac{C_b}{d}\right) \qquad [2.35]$$

Most of the authors used these dimensionless numbers to characterize the configuration of such a system. To be able to compare results with those of the literature, it is interesting (without it being compulsory!) to use this combination of internal measures.

2.4.5. From internal measures to the establishment of the process relationship

The dimensional analysis cannot go any further than equation [2.35]. In particular, it is unable to evaluate whether one or several internal measures have a significant effect on the target dimensionless number (N_P). The researcher must identify the function F, which is to say the mathematical form of the process relationship linking the internal measures relating to the causes to the power number (N_P).

To do this, an experimental program must be carried out in order to vary each internal measure representing the causes (Re, Fr, T/d, H_L/d, w/d and C_b/d) and to measure the results on the ratio $\pi_{target} = N_P$. If each internal measure varies independently from the others, it is obviously easier to analyze the impact of each one on N_P. This is why the choice of the repeated variables and rearrangements must also include, as far as possible, the experimental strategy. Although it is not clear at first glance, it is always possible to then carry out rearrangements between dimensionless numbers or to change the base. For instance, it is evident that, in order to independently vary the Froude number and the Reynolds number, it is necessary to rotate the impeller at the same speed but in fluids with different viscosities.

To establish the process relationship, Delaplace *et al.* [DEL 00, DEL 04] implemented a specific experiment program. We should specify straight away that it was purposefully carried out in tanks of extremely small dimensions. We will show in Chapter 5 that the process relationship established in this section can be applied *without any restriction* to larger tanks.

This experiment program is as follows:

– measures of consumed power ($0.05\ \mu W < P < 0.372$ W) are carried out during the stirring of viscous Newtonian fluids at various rotational speeds (1286×10^{-6} tr.s^{-1} $< N < 11.9$ tr.s^{-1});

– the Newtonian fluids are aqueous solutions of glucose with a concentration which is chosen so as to obtain different levels of viscosity and density (2.2 Pa.s $< \mu < 78.8$ Pa.s; 1360.5 kg.m^{-3} $< \rho < 1415.2$ kg.m^{-3}) at the temperature of the stirred media;

– different heights of liquid in the tank (0.024 m $< H_L < 0.082$ m) and different tank diameters (0.041 m $< T < 0.105$ m) are used;

– likewise, vertical plane impellers, with the same shape but of varying sizes (0.020 m $< d < 0.051$ m and 0.020 m $< w < 0.051$ m), are positioned at different heights from the bottom of the tank ($50\ 10^{-6}$ m $< C_b < 0.05$ m).

Since the *configuration* associated with this agitation process is described by six internal measures, a star graph is necessary in order to represent the evolution of the operating points obtained (Figure 2.11). To aid the clarity of the graph, not all operating points are shown, but it is nevertheless possible to view the range of variation covered by each internal measure individually. Froude and Reynolds numbers, respectively, cover 10 and six decades, but most geometric ratios only cover one, apart from C_b/d which extends over four orders of magnitudes. The predictive correlation which will be established will be valid over these ranges of variation.

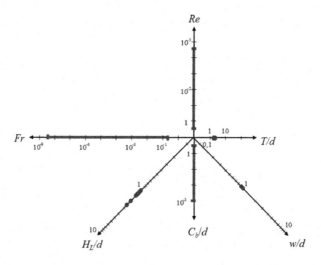

Figure 2.11. Guided example 1: graphic representation of the configuration of the system on a logarithmic scale (the range of values taken by each internal measure π_j during the experiment is shown by discrete points or gray bars with limits at each end)

To correlate our target internal measure (N_P), the mathematical form in a single monomial (see 2.2.1, equation [2.27]) is considered:

$$N_p = C\left(Re\right)^{e_2}\left(Fr\right)^{e_3}\left(\frac{T}{d}\right)^{e_4}\left(\frac{w}{d}\right)^{e_5}\left(\frac{H_L}{d}\right)^{e_6}\left(\frac{C_b}{d}\right)^{e_7} \qquad [2.36]$$

where C, e_2, e_3, e_4, e_5, e_6 and e_7, are the constants to be identified.

Before identifying the constants in equation [2.36], it is necessary to take into account certain constraints as defined in section 2.3.3. Since the dimensionless number w/d varied very little during the experiments (Figure 2.11), the influence of w/d on N_P cannot be properly verified, and therefore must be included in the constant C. Exponent e_5 is, therefore, forced to zero. This constraint leads us to reduce the configuration of the system to five internal measures. Equation [2.36] thus becomes:

$$N_p = C(Re)^{e_2}(Fr)^{e_3}\left(\frac{T}{d}\right)^{e_4}\left(\frac{H_L}{d}\right)^{e_6}\left(\frac{C_b}{d}\right)^{e_7}$$

[2.37]

By minimizing the sum of the squares of the differences between the experimental values of N_P and those predicted using equation [2.37], the following process relationship is obtained:

$$N_p = 199.4(Re)^{-1.017}(Fr)^{0.004}\left(\frac{T}{d}\right)^{-0.963}\left(\frac{H_L}{d}\right)^{0.103}\left(\frac{C_b}{d}\right)^{-0.048}$$

[2.38]

The mean error is lower than 4.2% for all of the operating points and never exceeds ±15%. It is interesting to note that the exponents linked to the Reynolds and Froude numbers are very close to those commonly observed in the literature, namely:

– an exponent close to −1 on the Reynolds number. Indeed, a given agitation has a power constant K_p in the laminar regime which characterizes its resistance to flow, and which is defined in the absence of a vortex formation during the agitation of a Newtonian fluid, whereby:

$$N_p = \frac{K_p}{Re}$$

[2.39]

– an exponent close to 0 on the Froude number, indicating therefore that the effects of gravity are negligible (the free surface of the stirred fluid remains perfectly flat).

By fixing the exponents of the Reynolds and Froude numbers at −1 and 0, respectively, equation [2.38] becomes (mean error of the order of 4.70%):

$$N_p = \frac{205.5}{Re}\left(\frac{T}{d}\right)^{-0.960}\left(\frac{H_L}{d}\right)^{0.076}\left(\frac{C_b}{d}\right)^{-0.048}$$

[2.40]

Therefore, if a plane impeller, vertically centered in a flat-bottomed tank, of any size and whose geometric ratios satisfy the conditions {1.31 < T/d < 2.59; 0.96 < w/d < 1; 1.17 < H_L/d < 2.68; 10^{-3} < C_b/d < 2.44}, puts into movement a Newtonian fluid in the conditions $\{4.37 \times 10^{-5} < Re < 2.47\,;\, 6.85 \times 10^{-9} < Fr < 0.30\}$, then the power number N_P can be predicted by equation [2.40], and thus the power to apply to the engine used to drive the impeller also.

2.4.6. Comments

COMMENT 2.1.– For the range of operating points tested, the value of the Froude number has no influence on N_P. However, there is nothing to say that this is the case for a different range. Indeed, the state-of-the-art shows that, for values of Reynolds numbers ($Re > 200$) and higher Froude numbers, it is not possible to neglect the influence of the Froude number on the power consumed. This clearly implies that, depending on the operating point being considered, the physical phenomena are no longer necessarily the same. When the Froude number (ratio between inertial and gravitational forces) becomes larger ($Fr \gg 1$), the free surface of the flow domain undergoes tangential and axial deformations, which may lead, depending on the rotational speed and the size of the tank, to the formation of a vortex around the stirring rod. If this vortex is present, it will modify the flow of the fluid in the tank, and therefore the consumption of power used to drive the impeller.

In conclusion, scientific literature dealing with the dimensional analysis of power for various mixing systems tells us that, for a fixed geometry, $N_P = F(Re, Fr)$, and allows us to identify various regimes:

– when $Re < 300$, the process relationship is reduced to a dependence of $N_P = F(Re)$ as made explicit in equation [2.40];

– when $Re > 300$ and the value of the Froude number is sufficiently raised so that a vortex is formed in the tank, the relationship is $N_P = F(Re, Fr)$;

– when $Re > 300$ and the value of the Froude number remains low (no vortex in the tank), the relationship is also $N_P = F(Re)$.

COMMENT 2.2.– The proximity of the stirrer to the side wall of the tank is given by the dimensionless number T/d. For the experimental program implemented, this ratio varies between 1.31 and 2.59. As a result, the space between the side wall of the tank and the stirrer remains relatively significant, thus avoiding any edge-effects between the tank and the stirrer. It is very likely that if this correlation were tested for a system where $T/d \approx 1$, it would not be satisfactory since (1) this value excludes the variation range of internal measures from which the process relationship was established and (2) the physical phenomena in play may differ.

COMMENT 2.3.– According to equation [2.40], the values of the exponents associated with both shape factors C_b/d and H_l/d are low (0.076 and –0.048, respectively). This shows that these internal measures, which characterize the geometric configuration of the system, do not have a strong influence on the target variable N_p. This is not the case for the shape factor T/d to which is associated an exponent equal to –0.96.

2.4.7. Conclusion

The previous discussion regarding the relevance of the correlation established to predict the power number shows:

– that it is essential to specify the range of validity of a process relationship;

– that such correlations can only be applied when the configuration of the system remains unchanged and the new operating point is included in the validity range of the process relationship.

3

Practical Tools for Undertaking the Dimensional Analysis Process

This chapter addresses certain points raised in Chapter 2, where the principles and methods for carrying out a dimensional analysis were discussed. It aims to highlight the difficulties which users may encounter while undertaking the dimensional analysis process themselves. This involves:

– facilitating the process of establishing the list of relevant physical quantities which influence the target variable and helping to evaluate the consequences of an omission/error in the compilation of this list (section 3.1),

– helping to make a suitable choice of the bases constituting the set of repeated physical quantities used as references (section 3.2),

– illustrating how to obtain a reduced configuration of the system (section 3.3) without restricting the range of validity of the process relationship.

For clarity purposes, this chapter consists mainly of examples. It is therefore a link between Chapter 2 (explanation of the dimensional analysis process for modeling) and Chapter 6 (focusing to case studies taken from the authors' research).

3.1. Establishment of the list of relevant physical quantities which influence the target variable

It is important to keep in mind that establishing the list of relevant independent physical quantities which influence the target variable is the most difficult stage of dimensional analysis because there is no predefined rule to govern this process. During this phase, the knowledge of the physical phenomena studied and experience in the modeling process are the most important criteria. Nevertheless, users can find some practical tools in this part which:

– allow them to adopt a generic/systematic approach when establishing this list;

– and illustrate the possible consequences of an omission of a physical quantity or the introduction of a superfluous physical quantity.

3.1.1. Approach to adopt in order to facilitate the establishment of the list of relevant physical quantities

As mentioned in Chapter 2 (section 2.1.3.1), the first condition to respect during the establishment of the list of relevant physical quantities influencing the target variable is the *physical independence of these quantities*. It is also advisable to list all the physical quantities which act on the phenomenon being studied, whatever their potential degree of influence.

Once these principles are in place, it becomes clear that the physical quantities which influence the target variable can be classified into four categories, regardless of the problem studied[1]:

– *material parameters*;

– *process parameters*;

– *boundary and initial conditions*;

– *universal constants*.

In order to make the establishment of the list of relevant physical quantities systematic, we advise methodically searching for these different types of physical quantities for any type of phenomenon being studied. Hereafter, we have chosen to illustrate our assertions with examples taken from food process engineering.

---

1 This classification is directly linked to the notion of the configuration of the system, as defined in Chapter 2 (section 2.3.1).

3.1.1.1. *Material parameters*

The material parameters are essentially linked to the characteristics of each of the phases present within a static reactor or the flow domain.

For a *single phase system*, it is generally not necessary to distinguish the properties of different components of the system. For a *multi-phase system*, it is necessary to distinguish:

– the material parameters of *the continuous phase*;

– the material parameters of *the dispersed phase*;

– the interfacial properties.

For instance, we can cite liquid–liquid systems such as mayonnaise containing two non-miscible fluids (e.g. oil and water), gas–liquid systems such as a foam, solid–liquid systems such as yoghurts with fruit pieces, or multi-phase systems containing all of these phases, such as ice-cream.

The material parameters correspond with the physical quantities which characterize the phases (or components present) in the conservation equations of mass, momentum and energy. Apart from some specific cases, experience shows that we often find:

– the density of the fluids or the solid particles, whatever the system;

– in the isothermal processes involving a flow of matter or molecular diffusion of components, the *diffusion coefficient*[2], the *rheological properties* of liquids or *viscoelastic* properties of solids, the *viscosity* of gases, the *solubility* of a component, and so on;

– the properties characterizing the capacity of the fluid to transfer heat, such as *specific heat* or *thermal conductivity*, and/or to conduct electricity, such as *electrical conductivity* in the non-isothermal processes;

– in the processes involving chemical reactions, reaction parameters such as the *activation energy*, the *rate constants of reaction, reaction orders*, and so on.

---

2 For interfacial multi-component mass transfers, it is appropriate to list the diffusion coefficients of each component.

The material parameters are not only restricted to the physical properties of the phases present. They also include the *characteristic factors which define the dispersed system* (e.g. diameter of the bubble, droplet or solid particle), the *mass or volume fractions* of the phases, the *void fraction*, and so on. It should be noted that these final two quantities are, by definition, dimensionless.

Finally, it is important to verify that *no algebraic relationship exists between the different material parameters listed*. For example, the kinematic viscosity v must not be added to the relevance list if dynamic viscosity μ and density ρ are already listed, because these three quantities are linked by the following relationship:

$$v = \frac{\mu}{\rho}$$

[3.1]

Another well-known example is thermal diffusivity[3] (α) which is defined using the thermal conductivity (λ), the density (ρ) and the specific heat (c_p) of the fluid or the solid:

$$\alpha = \frac{\lambda}{\rho.c_p}$$

[3.2]

3.1.1.2. *Process parameters*

In general, these are levers of the process, which is to say that they are control parameters of the process. For instance, this may involve the fluid flow rate in a continuous process, rotational impeller speed rate in a batch process, a given electrical power, and so on. If the process is not in steady-state, the duration of the process must also be listed.

First, we recommend that the reader only lists the process parameters which are directly measureable, and not the parameters deduced from the measured quantities. For example, list the volumetric flow rate of the liquid Q rather than the average fluid velocity $v = Q/A$ (where A is the cross-

---

3 Thermal diffusivity, just like kinematic viscosity, is in fact an intermediate variable. We will refer back to this concept in section 3.3. We will see how this type of variable helps to reduce the set of internal measures (configuration of the system) without restricting the range of validity of the process relationship.

sectional area of the cylindrical duct which the fluid passes), the impeller rotational speed N expresses in s^{-1} rather than the velocity at the impeller tip $\pi.N.d$ (where d is the diameter of the impeller). This approach attempts to ensure that the system is not oversimplified and that no physical quantities of interest are omitted. An average velocity is relevant for characterizing the hydrodynamics if the passage section is constant and the system monophasic. For more complex cases, Q/A, which is in reality an *intermediate variable* (see section 3), will be a truncated indicator for describing the distribution of the flow.

3.1.1.3. *Boundary and initial conditions.*

The term "boundary conditions" includes:

– the physical quantities imposed along the boundary of the system; for example, the temperature at the wall of a heat exchanger.

– the geometric parameters which define the flow domain. This is usually the volume or height of the fluid contained in a stirred tank, the diameter of a pipe, the clearance height which separates the impeller from the bottom of the tank, the mass of solids, and so on.

In practice, when the geometry of the system becomes too complex, it cannot be fully described by an acceptable number of geometric parameters. Only a restricted set of these parameters is listed in this case, chosen as a set of parameters representative of the size of the equipment. Nevertheless, it is important to bear in mind that this restricted set of geometric parameters does not fully describe the geometry of the system, and could therefore tacitly influence the target variable. As a consequence, we will call $\{p_{geo}\}$ *the set of geometric parameters other than the restricted set listed, necessary for the complete description of the flow domain.* With this set $\{p_{geo}\}$ will be associated (after transformation of the core matrix into an identity matrix) a set of *internal geometric measures* $\{\pi_{geo}\}\{\pi_{geo}\}$, which are generally shape factors of the system and/or the dimensionless numbers introduced in $\{p_{geo}\}$ (for example, the number of ribbons on a helical ribbon impeller). The fact that $\{\pi_{geo}\}$ is not explicitly expressed makes it more difficult to ascertain that the configuration of the system is identical on both scales.

In addition to these boundary conditions, the "*initial conditions*" should also be listed. With this term, we are referring to the physical quantities which define the system in its initial state. For example, we may cite the initial temperature of liquid media in a tank heated by a coil, or the initial concentration of a component.

3.1.1.4. *Special case: the universal physical constants*

The list of relevant physical quantities which influence the target variable occasionally contains *universal physical constants*. The most frequent of these universal constants is the gravitational acceleration g. Taking these constants into account is perfectly logical, in so far as they contribute to the definition of the system being studied. For instance, g defines the intensity of the gravitational field acting on the free surface of a liquid in a stirred tank. Therefore, g appears since this constant helps to define a boundary condition[4] of the system being studied.

Among the other universal physical constants in chemical engineering, we can cite Avogadro's number, the Boltzmann constant, the velocity of light in a vacuum, or Planck's constant.

3.1.2. *Consequences of omitting physical quantities*

It is possible that a decisive physical quantity may be omitted during the establishment of the list of relevant physical quantities which influence the target variable. This oversight leads to *an incomplete set of dimensionless numbers {π_i}*, since a dimensionless number π_j is missing in the description of the configuration of the system. This therefore gives *a partial, and thus incorrect, view of the causes governing the phenomenon being studied.*

We will illustrate this with the help of an example.

EXAMPLE 3.1.– Power consumed by a Newtonian fluid in a mechanically stirred tank (I)

We refer here to guided example no. 1 in Chapter 2 regarding the power consumed P by a Newtonian fluid in a mechanically stirred tank.

Imagine that the acceleration of gravity g had been accidentally omitted. Instead of equation [2.32], the problem would be translated in terms of independent physical quantities with the equation:

$$P = f\ (T, d, C_b, H_L, w, \mu, \rho, N) \tag{3.3}$$

---

4 Therefore, g could also be listed in the category of boundary conditions.

The power consumed (P) thus becomes the consequence of the eight physical quantities making up the second member of equation [3.3].

As in Chapter 2, consider the following repeated variables: ρ, N and d. The dimensional matrix associated with this problem is written in Chapter 2 (Figure 2.9), but this time, the column which refers to g must be concealed. After transformation of the core matrix into an identity matrix, the modified dimensional matrix is obtained where g is omitted from the relevance list, as represented in Figure 3.1.

	P	μ	////	T	w	H_L	C_b	ρ	N	D
M	1	1	0	0	0	0	0	1	0	0
-T	3	1	2	0	0	0	0	0	1	0
L+3M	5	2	1	1	1	1	1	0	0	1

Figure 3.1. *Example 3.1: modified dimensional matrix $\mathbf{D_m}$ (black outline) obtained by transforming the core matrix into an identity matrix where the physical quantity g has been omitted from the relevance list*

Using the coefficients contained in the modified residual matrix $\mathbf{R_m}$ presented in Figure 3.1, the six new dimensionless numbers are formed. Equation [2.33] then becomes:

$$\pi_{target} = \frac{P}{\rho . N^3 . d^5} = F_1\left(\frac{\mu}{\rho . N . d^2}, \frac{T}{d}, \frac{w}{d}, \frac{H_L}{d}, \frac{C_b}{d}\right) \qquad [3.4]$$

So, after rearrangements of internal measures (see equation [2.35]):

$$\pi_{target} = N_p = F_2\left(Re, \frac{T}{d}, \frac{w}{d}, \frac{H_L}{d}, \frac{C_b}{d}\right) \qquad [3.5]$$

In other words, the causes for the variation in the power number N_P (the target internal measure) are assimilated into five internal measures (here, R_e, $\frac{T}{d}$, $\frac{w}{d}$, $\frac{H}{d}$, $\frac{C_b}{d}$) although they are only fully described by a combination of six internal measures (Re, Fr, $\frac{T}{d}$, $\frac{w}{d}$, $\frac{H}{d}$, $\frac{C_b}{d}$), as shown in Chapter 2.

In order to illustrate the consequences of this truncated configuration, consider two fluids, 1 and 2, with different properties such as $\mu_2 = 100.\mu_1$ and $\rho_1 = \rho_2$. These two fluids are added to a single tank ($\dfrac{T}{d}$, $\dfrac{w}{d}$, $\dfrac{H}{d}$, $\dfrac{C_b}{d}$ fixed), and stirred at two different impeller rotational speeds ($N_2 = 100 \cdot N_1$). When the gravitational acceleration g is omitted from the relevance list, the operating point in cases 1 (μ_1, ρ_1, N_1) and 2 (μ_2, ρ_2, N_2) (μ_2, ρ_2, N_2) is identical since it is described by:

$$\left\{ Re_1 = \frac{\rho.N_1.d^2}{\mu_1} = \frac{\rho.(N_2/100).d^2}{(\mu_2/100)} = Re_2, \frac{T}{d}, \frac{w}{d}, \frac{H_L}{d}, \frac{C_b}{d} \right\} \qquad [3.6]$$

Then, we would expect to obtain the same value of the target internal measure, N_P. In fact, when writing the complete configuration of the system, we can observe that these two are not equivalent:

Case 1:
$$\left\{ \begin{array}{c} Re_1 = \dfrac{\rho.N_1.d^2}{\mu_1}, Fr_1 = \dfrac{N_1^2.d}{g}, \\[2ex] \dfrac{T}{d}, \dfrac{w}{d}, \dfrac{H_L}{d}, \dfrac{C_b}{d} \end{array} \right\} \qquad [3.7]$$

Case 2:
$$\left\{ \begin{array}{c} Re_2 = Re_1, Fr_2 = \dfrac{(100.N_1)^2.d}{g} = 10^4.Fr_1, \\[2ex] \dfrac{T}{d}, \dfrac{w}{d}, \dfrac{H_L}{d}, \dfrac{C_b}{d} \end{array} \right\} \qquad [3.8]$$

Equations [3.7] and [3.8] show that all the internal measures are conserved between case 1 and case 2, apart from the Froude number. This dimensionless number, which is an internal measure associated with gravitational acceleration (deformation of the free surface), varies by a factor of 10^4 from one case to another. Consequently, the operating points are different in the two cases: it is therefore highly likely that the value of the target internal measure may vary between these two cases. Typically, a vortex could be formed in case 2 and not in case 1, thus leading to a change in the order of magnitude of the power consumed.

3.1.3. *Consequences of introducing a superfluous physical quantity*

The introduction of physical quantities which play no role in the phenomenon being studied must be avoided since it generates an unnecessary enlargement of the configuration of the system. Consequently, the number of internal measures involved in the process relationship will be unhelpfully increased.

Moreover, special care should be taken not to list:

– *physical quantities which can be deduced using quantities already listed* (for example, kinematic viscosity, if ρ and μ have been listed). Indeed, in this case, the condition of physical independence between physical quantities is violated;

– *physical quantities dependent on other listed quantities, apart from the target variable.* For example, consider an agitation process in a stirred tank, with mixing time as the target variable. In this case, it is not possible to simultaneously consider among the causes influencing the target variable, the power dissipated in the stirred solution and the impeller rotational speed of the impeller. Indeed, both of these levers or commands induce the rotation of the impeller.

3.1.4. *Conclusion*

In conclusion, we should note that:

– Even though certain physical quantities (especially universal constants) do not vary during the study, they should not be systematically eliminated from the list of relevant physical quantities influencing the target variable. Indeed, combined with other physical quantities which vary during the process, they can lead to the different operating point.

– Operating points which appear identical may not lead to the same value of the target internal measurement if the list of relevant physical quantities used to build them is incomplete.

– The knowledge of the physical phenomena involved and the physical quantities which are involved in the associated physical equations significantly helps to establish the list of relevant physical quantities.

Finally, it should be pointed out that in some cases, *an incorrect configuration due to omitted or excessive quantities can be detected very early on*. Indeed, when the list of relevant physical quantities does not include at least two physical quantities covering the same fundamental dimension, it is not possible to obtain the complete set of dimensionless numbers: the modified residual matrix is therefore made up of a row only containing zero coefficients.

3.2. Choosing the base

As mentioned in Chapter 2, the set of repeated variables must comply with two conditions:

– they must be dimensionally independent;

– they must be chosen so as to express all of the dimensions of the physical quantities listed.

If, and only if, these two conditions are fulfilled, the number of repeated physical variables is equal to the number of fundamental dimensions required to express the dimensions of the physical quantities listed.

However, this set of repeated variables is not unique.

Generally speaking, we recommend:

– *that the target variable should not be chosen as a repeated variable*;

– *not choosing the physical quantities which may have a significant influence on the target variable as repeated variables*;

– *choosing the repeated variables so that the identification of the impact of each physical quantity on the target variable is facilitated*. Practically speaking, it is important to try to only give rise in an internal measure to a dimensional physical quantity which can be experimentally modified.

The objective of this section is to illustrate these assertions by referring to example 3.1, and in particular to demonstrate that:

– the phenomenological meaning of certain internal measures is not always immediately accessible, notably when they are not internal measures of fluxes (mass, momentum and energy), internal measures of retention or geometric internal measures (see Appendix 2).

– some $\{\pi_j\}$ sets may facilitate the determination of the relationship linking them to π_{target}.

EXAMPLE 3.1.– Power consumed by a Newtonian fluid in a mechanically stirred tank (II)

In guided example 1 (Chapter 2), we selected as a set of repeated variables, the base (noted here as base 1) made up of ρ, N and d. We will note $\{\pi_{i,base\ 1}\}$ the set of dimensionless numbers obtained with this base without carrying out rearrangements between numbers. It is expressed by:

$$\{\pi_{i,base1}\} = \begin{cases} \pi_{1,base1} = N_P = \dfrac{P}{\rho.N^3.d^5}, \pi_{2,base1} = \dfrac{\mu}{\rho.N.d^2}, \\[2mm] \pi_{3,base1} = \dfrac{g}{N^2.d}, \pi_{4,base1} = \dfrac{T}{d}, \\[2mm] \pi_{5,base1} = \dfrac{w}{d}, \pi_{6,base1} = \dfrac{H_L}{d}, \pi_{7,base1} = \dfrac{C_b}{d} \end{cases} \qquad [3.9]$$

Let us choose another base, denoted as base 2, defined by the set of repeated variables: P, ρ and μ.

The new modified dimensional matrix is shown in Figure 3.2.

	g	T	w	H_L	C_b	N	d	P	ρ	μ
-3M+L-2T	3	-1	-1	-1	-1	2	-1	1	0	0
-5M-2L-5T	4	-2	-2	-2	-2	3	-2	0	1	0
9M+3L+5T	-7	3	3	3	3	-5	3	0	0	1

Figure 3.2. *Example 3.1: creation of the modified dimensional matrix* $\mathbf{D_m}$ *(black outline) by choosing as repeated variables P, ρ and μ (noted base 2)*

From the coefficients contained in the modified residual matrix defined in Figure 3.2, the set of dimensionless numbers associated with this base 2 (P, ρ, μ) can be obtained. This set, denoted as $\{\pi_{i,base\ 2}\}$, is as follows:

$$\left\{ \pi_{i,base\,2} \right\} =$$

$$\left\{ \begin{array}{l} \pi_{1,base\,2} = \dfrac{g.\mu^7}{P^3.\rho^4} \,, \pi_{2,base\,2} = \dfrac{T.P.\rho^2}{\mu^3} \,, \pi_{3,base\,2} = \dfrac{w.P.\rho^2}{\mu^3} \,, \\[3mm] \pi_{4,base\,2} = \dfrac{H_L.P.\rho^2}{\mu^3} \,, \pi_{5,base\,2} = \dfrac{C_b.P.\rho^2}{\mu^3} \,, \pi_{6,base\,2} = \dfrac{N.\mu^5}{P^2.\rho^3} \,, \\[3mm] \pi_{7,base\,2} = \dfrac{d.P.\rho^2}{\mu^3} \end{array} \right\} \qquad [3.10]$$

Therefore, by choosing as repeated variables (P, ρ and μ), seven dimensionless numbers $\{\pi_{i,base\,2}\}$ describe this problem, as much as the numbers obtained previously $\{\pi_{i,base\,1}\}$. As defined in equation [3.10], these are respectively the internal measures of the physical quantities g, T, w, H_L, C_b, N and d.

It is not easy to immediately understand the meaning of the dimensionless numbers $\pi_{2,\,base2}$ to $\pi_{5,\,base2}$. However, when dividing each one by the dimensionless number $\pi_{7,\,base2}$, it is possible to bring up the shape factors of geometric quantities of $\{\pi_{i,\,base\,1}\}$: $\dfrac{T}{d}$, $\dfrac{w}{d}$, $\dfrac{H_L}{d}$, $\dfrac{C_b}{d}$.

Therefore, the space $\{\pi_{i,\,base\,2}\}$ can be recombined to give space $\{\pi'_{i,\,base\,2}\}$ as follows:

$$\left\{ \pi'_{i,base\,2} \right\} =$$

$$\left\{ \begin{array}{l} \pi'_{1,base\,2} = \dfrac{g.\mu^7}{P^3.\rho^4} \,, \pi'_{2,base\,2} = \dfrac{T}{d} \,, \pi'_{3,base\,2} = \dfrac{w}{d} \,, \pi'_{4,base\,2} = \dfrac{H_L}{d} \,, \\[3mm] \pi'_{5,base\,2} = \dfrac{C_b}{d} \,, \pi'_{6,base\,2} = \dfrac{N.\mu^5}{P^2.\rho^3} \,, \pi'_{7,base2} = \dfrac{d.P.\rho^2}{\mu^3} \end{array} \right\} \qquad [3.11]$$

Written in this way, it is clearer that the power consumed P depends, among other things, on purely geometric internal measures. This is why it is possible to say that the rearrangement leading to a new space $\{\pi'_{i,\,base\,2}\}$ is relevant for describing the configuration of the system.

However, in this space $\{\pi'_{i,\,base\,2}\}$, the target variable (power P) appears again in several internal measures. In practice, this means that the researcher will inevitably modify the power P by varying the impeller rotational speed N and, as a consequence, the three internal measures $\pi'_{1,\,base\,2}$, $\pi'_{6,\,base\,2}$ and $\pi'_{7,\,base\,2}$ which are dependent on it. This undesired result could have been predicted since P is part of the base. This confirms how important it is not to choose the target variable as a repeated variable, as stated earlier. Rearrangements of dimensionless numbers must therefore be carried out so that each system control variable (i.e. variable which may be modified in the experimental program) is only present in one internal measure. By replacing $\pi'_{1, base2}$ with $(\pi'_{1, base 2}).(\pi'_{7, base 2})^3$, and $\pi'_{6, base2}$ with $(\pi'_{6, base 2}).(\pi'_{7, base 2})^2$, the space $\{\pi'_{i, base 2}\}$ is transformed into a space $\{\pi''_{i, base 2}\}$ whereby:

$$\left\{ \pi''_{i,base2} \right\} = \left\{ \begin{array}{l} \pi''_{1,base2} = \dfrac{\rho^2.g.d^3}{\mu^2}, \; \pi''_{2,base2} = \dfrac{T}{d}, \; \pi''_{3,base2} = \dfrac{w}{d}, \; \pi''_{4,base2} = \dfrac{H_L}{d}, \\[3mm] \pi''_{5,base2} = \dfrac{C_b}{d}, \; \pi''_{6,base2} = \dfrac{\rho.N.d^2}{\mu}, \; \pi''_{7,base2} = \dfrac{d.P.\rho^2}{\mu^3} \end{array} \right\} \quad [3.12]$$

It should be noted that this set could have been obtained directly if base 3 made up of repeated variables (d, ρ and μ) had been initially chosen to establish the relationship process. The dimensionless numbers $\pi''_{1, base 2}$ to $\pi''_{6, base 2}$, respectively, provide internal measurements of the physical quantities g, T, w, H_L, C_b and N. They therefore represent the causes responsible for the changes in the target internal measure, $\pi''_{7, base 2}$. This time, when the researcher varies the impeller rotational speed N, only the dimensionless number $\pi''_{6, base 2}$ (which is nothing but a Reynolds number) will be impacted , apart from the target variable $\pi''_{7, base 2}$, of course.

3.3. Some techniques for reducing the set of dimensionless numbers (configuration of the system)

This section looks to go further in the dimensional analysis process by showing that certain techniques can *reduce the configuration of the system* (the set of internal measures responsible for the variation of the target internal measure), without *restricting the range of validity of the process relationship*.

Such a reduction offers the advantage of:

– providing process relationships whose analytical expression and graphical representation are simplified;

– reducing the experiment program from which the parameters for the process relationship are determined;

– facilitating the physical interpretation of internal measures and their influence on the target internal measure.

According to the Vaschy–Buckingham theorem, any homogeneous relationship between m independent quantities measured by n_d fundamental dimensions can be represented in the form of another relationship between $(m$-$n_d)$ dimensionless numbers. As a consequence, a reduction in the configuration of the system (the number $(m$-$n_d)$ of internal) *without restricting the range* of *validity* of the process relationship can be done:

– either by reducing the number of physical quantities listed (m); or

– by increasing the number of fundamental dimensions (n_d); or

– by combining both of these approaches.

To this end, three techniques are available:

– the introduction of an intermediate variable instead of several physical quantities;

– the splitting of a fundamental dimension into several dimensions of the same nature;

– the introduction of a new fundamental dimension.

Each of these techniques is illustrated below by means of examples. We would like to warn the user of the fact that the use of these techniques (and especially the last two), *as attractive as they may seem, requires an in-depth knowledge of the physical phenomena involved. If this is not the case, it can lead to an incorrect configuration of the system. Extreme caution is therefore recommended when these techniques are implemented.*

3.3.1. *Introduction of an intermediate variable*

In many cases, physical laws show that the influence of various physical quantities on the phenomenon being studied must not be considered in an isolated way (i.e. independently from each other). On the contrary, they must be considered in a combined way by grouping them in a third variable through an analytical expression.

When several physical quantities listed are replaced with a substitute variable expressed as a combination of these different quantities, an *intermediate variable,* as is denoted in this book, is introduced into the dimensional analysis. The use of an intermediate variable helps to reduce the configuration of the system, without restricting the range of validity of the process relationship.

The advantages and limitations of this technique are highlighted in the example below, in the light of the physical phenomena involved. The objective is to warn the user of the *risks of systematically using intermediate variables without careful reflection.*

Finally, the user is invited to refer to the fourth example presented in Chapter 6 (powder mixing) in which this technique is implemented.

EXAMPLE 3.2.– Terminal velocity of a bead falling in a viscous liquid at rest (I)

In this example, the target variable is the terminal velocity v of a bead falling in a Newtonian liquid at rest and theoretically infinite so that the effects of the tank walls can be disregarded. We can assume that the liquid is viscous enough to ignore the inertial forces when compared to the viscous forces. In this case, when the bead is released, its falling velocity grows until the buoyancy and drag forces match the weight of the bead. At this stage, acceleration becomes zero and terminal velocity v is reached. Physical quantities which have an influence on v are as follows:

– the *material parameters*: these are the densities of the bead, ρ_b, and of the surrounding liquid, ρ, as well as the dynamic viscosity of the liquid, μ;

– the *process parameters*: there is none, since the liquid is at rest and the bead is released at an initial velocity of zero;

– *boundary and initial conditions*: they are reduced to a single geometric parameter which is the diameter of the bead d (infinite environment);

– *universal constant*: in this case, gravitational acceleration g.

Therefore, this problem can be reduced to six physical quantities whose dimensions are expressed using three fundamental dimensions (M, L and T). A process relationship between three numbers will therefore be expected. The following modified dimensional matrix is obtained (Figure 3.3), by taking the repeated variables as (ρ, μ, g).

	v	ρ_b	d	ρ	μ	g
-M-2/3L-1/3T	-1/3	1	-2/3	1	0	0
2M+2/3L+1/3T	1/3	0	2/3	0	1	0
-M-1/3L-2/3T	1/3	0	-1/3	0	0	1

Figure 3.3. *Example 3.2: modified dimensional matrix* $\mathbf{D_m}$
(black outline). Identity matrix $\mathbf{C_I}$ *in dark gray, and the modified residual matrix* $\mathbf{R_m}$ *in light gray*

The coefficients contained in the modified residual matrix given in Figure 3.2 leads to the following relationship:

$$v.\sqrt[3]{\frac{\rho}{\mu.g}} = F\left(d.\sqrt[3]{\frac{g.\rho^2}{\mu^2}}, \frac{\rho_b}{\rho}\right) \qquad [3.13]$$

where F is the process relationship required.

Assuming that F can be approximated by a monomial function, whereby:

$$v.\sqrt[3]{\frac{\rho}{\mu.g}} = A_1.\left(d.\sqrt[3]{\frac{g.\rho^2}{\mu^2}}\right)^{b_1}.\left(\frac{\rho_b}{\rho}\right)^{b_2} \qquad [3.14]$$

where A_1, b_1 and b_2 are constants.

If the density of the bead ρ_b is equal to that of the surrounding liquid ρ, the bead will remain in static equilibrium and its velocity v will be zero. However, equation [3.14] does not yield this result, if $\frac{\rho_b}{\rho}=1$. This shows that equation [3.14] is not suitable for describing this problem and that the process relationship does not have a monomial form.

At this stage, it is necessary to go further in examining the physical phenomena involved when a bead falls in a liquid at rest with a zero initial velocity. Indeed, during the fall, the bead is subject to a force resulting from buoyancy and weight. Consequently, the influence of the gravitational acceleration and densities of the bead and liquid is inevitably combined into a single term: $\Delta\rho.g$. On the basis of this knowledge, $\Delta\rho.g$ can be introduced as an *intermediate variable*, instead of individually considering the densities of the bead and the surrounding fluid on the one hand, and gravitational acceleration on the other hand. The problem is reduced in this case to only three quantities of influence and the number of internal measures required is thus equal to 4-3 = 1: a process relationship in the form of a singleton is therefore expected. Given that it is advised not to consider the target variable as a repeated variable, the base chosen is made up of repeated variables d, $\Delta\rho.g$ and μ.

The new modified dimensional matrix is shown in Figure 3.4.

	v	d	$\Delta\rho.g$	μ
L-T	2	1	0	0
-T-M	1	0	1	0
2M+T	-1	0	0	1

Figure 3.4. *Example 3.2: modified dimensional matrix* **D**$_m$ *(black outline) in the case where the intermediate variable $\Delta\rho.g$ is introduced*

The coefficients contained in the modified residual matrix in Figure 3.4 lead to the relationship:

$$\frac{v.\mu}{d^2.\Delta\rho.g} = C_0 \qquad [3.15]$$

where C_0 is a constant. Equation [3.15] can also be written as:

$$v = C_0.\frac{d^2.\Delta\rho.g}{\mu} \qquad [3.16]$$

COMMENT 3.1.– This example confirms that the introduction of physical knowledge through an intermediate variable into the dimensional analysis process helps to advantageously reduce the configuration of the system and lead to a process relationship very close to the analytical law. Indeed, equation [3.16] is nothing but the familiar Stokes' law, where the constant C_0 is equal to 1/18.

COMMENT 3.2.– As previously mentioned, the relationship obtained can only be applied to very low particle Reynolds numbers (Re_p<<1, corresponding to a creeping flow) for which the viscous forces are dominant, which helps to list $\Delta\rho$ instead of the densities of the liquid and bead. However, for higher Reynolds numbers, the density of the liquid must be listed in addition to the intermediate variable $\Delta\rho.g$ to take account of the effects of inertia on the terminal velocity of the falling bead.

This highlights once again the need to have an in-depth knowledge of the physics of the phenomena present (ideally through analytical relationships) when an intermediate variable is introduced. This ensures that the impact of the physical quantities replaced in this way is properly taken into account by this new variable.

3.3.2. *Increase in the number of fundamental dimensions*

The reduction in the number of internal measures can also be obtained by increasing the number of fundamental dimensions. Two techniques are available to achieve this objective: splitting a fundamental dimension or introducing a new fundamental dimension. In any case, this means that *a new dimension is added to the set of fundamental dimensions associated with the International System of Units, thus leading to the definition of a new system of units.* It should be remembered that the only obligation to respect in order to carry out a dimensional analysis process is to use a *coherent system of units* (see Chapter 2, section 2.1.1.2.1).

3.3.2.1. *Splitting a fundamental dimension*

According to the Vaschy–Buckingham theorem, a reduction in the configuration of the system can be obtained by increasing the number of fundamental dimensions n_d. To do this, the first method involves *splitting a fundamental dimension into two fundamental dimensions of the same nature.*[5]

We will present this technique using an example from Szirtes [SZI 07], to which the splitting of the fundamental dimension of length [L] will be applied.

EXAMPLE 3.3.– Rising fluid in a capillary tube

Imagine a glass tube whose base is plunged into a liquid which is at rest. Due to capillary forces, the liquid will climb up the tube, the height of which is denoted as *h*. The system is then considered as being in equilibrium.

The target variable in this problem is *h* (Figure 3.5).

---

5 This approach is denoted as "dimension splitting" by Szirtes [SZI 07].

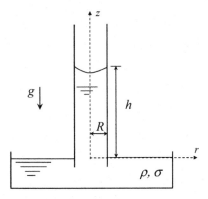

Figure 3.5. *Example 3.3: rising fluid in a capillary tube*

The physical quantities influencing the value of *h* are:

– the *material parameters*: which are the density of the liquid ρ and the surface tension of the liquid σ. The viscosity of the liquid μ must not be listed because the viscous forces do not apply when the system is in equilibrium, i.e. there is no flow;

– the *process parameters*: there are no quantities to be considered in this category;

– the *boundary and initial conditions*: the radius of the tube *R*. As the height of the liquid *h* in the tube is measured from the free surface of the vessel, it is not necessary to take account of the vessel's geometry or volume;

– *universal constants*: gravitational acceleration, *g*.

The physical problem can therefore be represented by a relationship between five physical quantities expressed using three fundamental dimensions: two internal measures are therefore required. The repeated variables chosen are *R*, *g* and σ. The modified dimensional matrix obtained is presented in Figure 3.6.

	h	ρ	R	g	σ
2M+2L+T	1	-2	1	0	0
-2M-T	0	-1	0	1	0
M	0	1	0	0	1

Figure 3.6. *Example 3.3: modified dimensional matrix* $\mathbf{D_m}$ *(black outline)*

The coefficients contained in the modified residual matrix lead to the following process relationship:

$$\frac{h}{R} = F\left(\frac{\rho.R^2.g}{\sigma}\right)$$ [3.17]

Once again, dimensional analysis does not allow us to go further: the mathematical form of the process relationship can only be provided by implementing an experimental program. However, it is possible to reduce the configuration of the system and simplify the previous process relationship by splitting a fundamental dimension. With regard to the physical phenomena involved, it is advisable to separate the fundamental dimension of length (L) into two fundamental dimensions:

– one dimension related to the axial length , denoted as $[L_z]$;

– the second dimension related to the radial length, denoted as $[L_r]$.

$[L_z]$ and $[L_r]$, respectively, relate to the vertical axis z and the radial axis r as defined in Figure 3.5.

It is then a case of expressing the dimensions of the physical quantities previously listed in this new system of fundamental dimensions:

– the gravitational acceleration g acts on the vertical axis z, the dimension of length considered to express its dimension should then be replaced with $[L_z]$;

– following the same logic, the height of the liquid in the tube h should be expressed as an altitude $[L_z]$;

– in this problem, a volume corresponds to the product of the square of a radial distance (a surface) and an altitude (an axial length: a height). Since the density of the liquid in the tube is defined with respect to a volume, its dimension $[M.L^{-3}]$ should be replaced with $[M.L_z^{-1}.L_r^{-2}]$;

– finally, the surface tension can be analyzed as a force being exerted vertically (opposite to gravitational force with a dimension of $[M.L_z.T^{-2}]$) on the edge of the meniscus with a dimension of $[L_r]$; consequently giving:

$$[\sigma] = [M.L_z.T^{-2}.L_r^{-1}]$$ [3.18]

Equation [3.18] shows that:

– even if the dimension of the surface tension does not explicitly involve the fundamental dimension of length L in the International System of Units (since we have $[\sigma] = [M.T^{-2}]$), its dimension in the new system of units is expressed using the two fundamental dimensions of length $[L_z]$ and $[L_r]$;

– moving from the International System of Units to the new system obtained by splitting the dimension of length does not give rise to numerical coefficients of proportionality other than one when the derived (or secondary) units are deduced from fundamental units (see Chapter 2, section 2.1.1.2.1). This new system is therefore coherent.

The problem is thus defined by five physical quantities and four fundamental dimensions of the new system of units. Therefore, a process relationship between 5-4 = 1 internal measure is expected, that is to say, a singleton. Since h is the target variable, the repeated variables chosen are ρ, R, g and σ. The new modified dimensional matrix is presented in Figure 3.7.

	h	ρ	R	g	σ
$-2\,L_z-T$	-1	1	0	0	0
$2M-2L_z+2\,L_r-T$	-1	0	1	0	0
$-M-L_z-T$	-1	0	0	1	0
$2M+2\,L_z+T$	1	0	0	0	1

Figure 3.7. *Example 3.3: modified dimensional matrix D_m after splitting a fundamental dimension (black outline)*

The process relationship is therefore restricted to a singleton equal to a constant C_0:

$$\frac{h.\rho.R.g}{\sigma} = C_0 \tag{3.19}$$

which can be written in the following form:

$$h = C_0.\frac{\sigma}{\rho.R.g} \tag{3.20}$$

This new form of the process relationship, known to within one constant, thus indicates that the height of the liquid in the capillary tube is inversely proportional to the radius of the tube. It is due to the splitting of the fundamental dimension of length that this information could be provided. This was not directly accessible using the relationship in equation [3.17]. It can be shown that equation [3.20] corresponds to Jurin's law in which the constant is equal to *2cos α,* where *α* is the contact angle between the liquid and the wall of the tube.

3.3.2.2. *Introducing a new fundamental dimension*

The second method for increasing the number of fundamental dimensions n_d consists of *introducing a new fundamental dimension.*

We will illustrate this by referring back to the example of the bead falling in a viscous liquid at rest (example 3.2).

EXAMPLE 3.2.– Terminal velocity of a bead falling in a viscous liquid at rest (II)

Let us analyze this example by introducing a new fundamental dimension, in addition to the seven fundamental dimensions used in the International System of Units. This is a fundamental dimension of force, denoted as [N], referring to the Newton.

Given that, in the International System of Units, the dimension of a force is $[M.L.T^{-2}]$, the dimension of the velocity can be expressed in this new system of units including force as a fundamental dimension, whereby:

$$[v] = \left[\frac{N}{M.L.T^{-2}} .L.T^{-1} \right] = \left[N.M^{-1}.T \right] \tag{3.21}$$

The same process can be used for acceleration of gravity, linked to the gravitational force:

$$[g] = \left[\frac{N}{M.L.T^{-2}} .L.T^{-2} \right] = \left[N.M^{-1} \right] \tag{3.22}$$

The other physical quantities, not directly linked to forces, are not expressed according to this new fundamental dimension, but only in accordance with M, L and T, as before.

The problem is now defined by five physical quantities and four fundamental dimensions of the new system of units. The set of repeated variables chosen is d, Δp, μ, g (v being the target variable). The number of internal measures is 5-4 = 1, and thus a singleton once again. The new modified dimensional matrix is presented in Figure 3.8.

	v	d	Δp	μ	g
3M+L+2T+3N	2	1	0	0	0
M+T+N	1	0	1	0	0
-T	-1	0	0	1	0
N	1	0	0	0	1

Figure 3.8. *Example 3.2: modified dimensional matrix* **D**m *(black outline) after introducing the fundamental dimension of force N*

The modified residual matrix described in Figure 3.8 leads to the process relationship:

$$\frac{v.\mu}{d^2.\Delta p.g} = C_0 \qquad\qquad [3.23]$$

where C_0 is a constant, whereby:

$$v = C_0.\frac{d^2.\Delta p.g}{\mu} \qquad\qquad [3.24]$$

It can be observed that the introduction of a new fundamental dimension helps to directly obtain the process relationship [3.16], identical to Stokes' law.

3.3.2.3. Conclusion

These examples demonstrate that, when a new system of units is defined, either by splitting a fundamental dimension, or by introducing a new fundamental dimension, the main difficulty is expressing the dimension of the physical quantities in the new system of units considered. Even more so for the use of an intermediate variable, *the increase in the number of fundamental dimensions requires very good knowledge of the physics of the phenomena associated with the case being studied.*

Dimensional Analysis of Processes Influenced by the Variability of Physical Properties

4.1. Introduction

Food processes involve matter (fluid, emulsion, suspension, etc.), which is placed in equipment in order to be transformed. In numerous cases, one (or several) physical properties of the matters change between the inlet and the outlet of the equipment. There is also very often a spatial and/or temporal distribution of these physical properties within the equipment insofar as they are dependent on various potential fields (composition, temperature, etc.). This dependence has a significant effect on the process and must be taken into account. For instance:

– the variation of viscosity with temperature which can be observed within an exchanger during thermal processing. This variation in viscosity with temperature modifies the heat transfer performance of the process when compared to a fluid whose thermo-dependence is negligible;

– the variation of apparent viscosity with shear rate for non-Newtonian fluids. The distribution of apparent viscosity significantly influences the velocity field within the equipment compared to that obtained when viscosity is constant (Newtonian fluids). For example, this is the case for the flow of fluids presenting a yield stress in a stirred tank: a formation of caverns[1] is thus observed, which does not occur during the agitation of Newtonian fluids.

It is also relevant to mention the dependence of most thermodynamic properties (diffusion coefficient, solubility, volatility, etc.) or physical properties (density, thermal conductivity, specific heat, surface tension, etc.) on temperature. The influence of temperature on viscosity is, in many heating or cooling processes, more significant than other properties. As a result, only the variability of this property in the process will be taken into account. However, some caveats should be made: for instance, when processes are guided by free convection, the thermo-dependence of density must be taken into account because it leads to convection currents within the fluid.

The rigorous modeling and optimization of the processes must inevitably take account of this variability of material physical properties, and understand its influence on the process. This is even more evident to address scale change issues between an industrial process (1/1 scale) and a laboratory pilot or cold prototype ($1/F_e$ scale). As will be shown in Chapter 5, the identity of the "operating point" on both scales makes it necessary to conserve the numerical values of the set of internal measures (dimensionless numbers), namely to ensure complete similarity between the industrial process and the laboratory equipment. This similarity includes geometric similarity, process similarity (mechanical, static, kinematic, dynamic, thermal, chemical, luminous similarities, etc.), as well as *material similarity*[2]. This notion is illustrated using two practical situations which a scientist may be faced with:

– a process relationship has been established by considering the physical properties of the product constant: is it possible to reasonably extend this relationship to another scale or to another product? If so, what are the limitations on this extension?

– in view of validating the feasibility of a thermal process at high temperature (1/1 scale) using a product whose property varies with temperature, experiments with an ambient temperature prototype

---

1 The term "cavern" refers to the area surrounding the impeller where there is a movement of fluid, even though, elsewhere in the tank, the fluid is at rest.
2 The notion of material similarity will be precisely defined in this chapter. Its consequences are dealt with in detail in Chapter 5.

($1/F_e$ scale) need to be conducted: what product models and/or temperature ranges should be chosen in the prototype working at ambient temperature?

To answer these questions, dimensional analysis is an effective tool. Widely applied in the case of constant physical properties, the implementation of dimensional analysis for a variable physical property unfortunately remains almost unexplored. This goes a long way in explaining why most studies in chemical engineering still consider constant physical properties in their dimensional analysis. The only exception is the dependence on temperature of Newtonian viscosity, which is traditionally taken into account by introducing a viscosity number Vi raised to a given exponent. This viscosity number Vi is thus defined as:

$$Vi = \frac{\mu_b}{\mu_w} \qquad\qquad [4.1]$$

where μ_b is the viscosity at the average bulk temperature and μ_w is the viscosity at the average wall temperature. For instance, in the correlation of Sieder and Tate [SIE 36] describing forced convection in a pipe in a laminar regime, the viscosity number is at an exponent of 0.14.

In the following section, we will discuss the conditions which should be fulfilled in order that the generalization of such an internal measure to other dimensional analysis models is permitted.

It should be noted that in the presence of a variable physical property, dimensional analysis as presented in the previous chapters cannot be conducted in as straightforward a way. Indeed, since the material physical property is not constant, it must be described by a function which we will call a *material function*, such as a rheological law. It will, therefore, be necessary to take account of this material function to establish the list of all the relevant physical quantities which characterize the process. In this chapter, we will show that, when a process involving a variable physical property is compared to the same process in which this property is considered as being constant:

– the number of internal measures necessary for the description increases;

– the process relationship describing the evolution of the target variable is altered.

4.1.1. *Objective*

The objective of this chapter is to show how to carry out a rigorous dimensional analysis in cases where the process is influenced by the variability of a physical property within it. This largely consists of:

– describing the theoretical framework and the tools available for establishing an unbiased relevance list including all the parameters which describe the variability of the physical property;

– showing that this theoretical framework complies with the principles of the theory of similarity and helps to define a procedure which can be applied and repeated whatever physical property is being considered.

This chapter is at the heart of writing this book and undoubtedly the prime motivation. We believe that it provides an original contribution compared to the existing literature which often avoids this issue. This is why we encourage the user to pay particularly close attention to it so that the application of dimensional analysis in the case of variable physical properties is no longer the reserve of a few insiders.

This chapter extensively uses the work of Pawlowski [PAW 91] which defined the rules to comply with in the case of variability of physical properties, but which was unfortunately not sufficiently disseminated to be applied on a large scale.

4.1.2. *Approach*

This chapter is structured into four main sections.

In section 4.2, we will provide several definitions (section 4.2.1): dimensional material function $s(p)$, dimensionless material function $H(u)$, argument u and reference abscissa p_0. We will then show that the process relationship depends on the dimensionless material function, and not on its dimensional form (section 4.2.2).

Section 4.3 is devoted to dimensionless material functions. Above all:

– we will show that the form of graphical representations of the dimensionless material function $H(u)$ depends on the dimensional material function $s(p)$, as well as on the reference abscissa p_0 and the non-dimensionalization method (section 4.3.1);

– we will show that not all the non-dimensionalization methods of the material function comply with the principles of the theory of similarity (section 4.3.2);

– we will present the standard non-dimensionalization method which is based on an approximation of the dimensional material function $s(p)$ (Taylor series) in the vicinity of the reference abscissa p_0, normalized by $s(p_0)$. This method notably helps to create a theoretical framework which complies with the principles of the theory of similarity (section 4.3.3), whatever the nature of the material function;

– we will present the reference-invariant dimensionless material function, which is a standard dimensionless material function presenting the characteristic that is to be independent on the reference abscissa. They can be described by very specific mathematical equations (section 4.3.4).

Section 4.4 examines the construction of a space $\{\pi_i\}$ which is compatible with the principles of the theory of similarity in the case of a material with a variable physical property in the process. This involves:

– providing the tools to determine the supplementary dimensionless numbers and/or physical quantities to introduce to the initial relevance list to take account of the variability of the physical property (section 4.4.1);

– illustrating this approach by using various examples of material functions (section 4.4.2);

– showing how to make a relevant choice of the reference abscissa p_0 (section 4.4.3).

Section 4.5 provides a guided example in which modeling by dimensional analysis is carried out in a process involving a variable physical property. It deals with the frictional pressure drop associated with the laminar flow of non-Newtonian fluids in a straight and smooth pipe of circular cross-sectional area.

The approach which we have adopted in this chapter combines complete mathematical demonstrations and generic properties sometimes only illustrated by examples. This compromise was born from a wish to demonstrate that dimensional analysis involving a material with a variable property uses a rigorous theoretical and mathematical framework, and to highlight the key elements in implementing this approach.

To supplement this chapter, the user should refer to:

– Chapter 5 in order to understand how the tools and theoretical framework introduced in this chapter must be used in order to address scale change issues related to processes involving a material with a variable physical property;

– Chapter 6 for the application of these tools on examples taken from our own research with a view to establish a process relationship.

4.2. Influence of the material function on the process relationship and material similarity

4.2.1. Definitions and notations

In this chapter, we will keep the formalism set out by Pawlowski [PAW 91]. We will, therefore, note:

– $s(p)$, the *dimensional material function* describing the variation in the physical property of the material s with variable p whereby:

$$p \mapsto s(p) \tag{4.2}$$

– *H(v)*, the dimensionless material function defined by:

$$v \mapsto H(v) = \frac{s(p)}{s(p_0)} \tag{4.3}$$

where $s(p_0)$ is the value of the dimensional material function at any reference abscissa denoted as p_0, and v is the *argument* of the dimensionless material function of variable p. By definition, *the argument v is dimensionless.*

We chose the term "material function" in reference to the terminology used by Pawlowski [PAW 91] and maintained by Zlokarnik [ZLO 06]. In this book, "material function" will only refer to a variable physical property of a material involved in a process. This material may be a fluid (or a mixture of fluids), a phase (or several phases), an agro-food product and so on.

EXAMPLE 4.1 – Pseudoplastic fluid (I)

The rheological law describing the variation in apparent viscosity μ_a (Pa.s) depending on the shear rate $\dot{\gamma}\,(s^{-1})$ is expressed for a pseudoplastic fluid as follows:

$$\mu_a = k.\dot{\gamma}^{n-1} \tag{4.4}$$

where k (Pa.s^n) and n (-) are the fluid's consistency and flow index, respectively.

In this case, $p = \dot{\gamma}$ and $s = \mu_a$, the dimensional material function is therefore:

$$s(p) = \mu_a(\dot{\gamma}) = k.\dot{\gamma}^{n-1} \tag{4.5}$$

The associated dimensionless material function, $H(v)$, is:

$$H(v) = \frac{\mu_a}{\mu_0} = \frac{k.\dot{\gamma}^{n-1}}{k.\dot{\gamma}_0^{n-1}} = \left(\frac{\dot{\gamma}}{\dot{\gamma}_0}\right)^{n-1} \tag{4.6}$$

where μ_0 is the apparent viscosity calculated at any reference shear rate $\dot{\gamma}_0 : \mu_0 = \mu_a(\dot{\gamma}_0)$.

By defining the argument v of the dimensionless material function H as:

$$v = \frac{\dot{\gamma}}{\dot{\gamma}_0} \tag{4.7}$$

This gives:

$$v = \frac{\dot{\gamma}}{\dot{\gamma}_0} \;\mapsto\; H(v) = v^{n-1} \tag{4.8}$$

4.2.2. Illustrating the influence of the variability of a physical property on the target internal measure

In this section, we will highlight the fact that the target variable and the process relationship which links it to other internal measures are altered in the presence of a material with a variable physical property.

To do so, we will examine the example of the pressure drop of a non-Newtonian fluid flowing through a circular, straight and smooth section of pipe (laminar regime).

The balance of the forces acting on an elementary ring of fluid located at a radius of r ($0 < r < R$) and with a thickness of dr can be established in the same way as for Newtonian fluids [MCC 93]. It is then shown that, when the regime is laminar, the axial component of velocity $v_z(r)$, defined at radial position r, satisfies the equation:

$$\frac{\tau_w r}{R} + \mu_a \cdot \frac{d v_z}{d r} = 0 \qquad [4.9]$$

and that $\dfrac{\tau}{\tau_w} = \dfrac{r}{R}$ $\qquad\qquad$ [4.10]

where R is the radius of the pipe, τ is the shear stress, τ_w is the wall shear stress and $\mu_a = \dfrac{\tau}{\dot{\gamma}}$ is the apparent viscosity of the fluid which in this case replaces the Newtonian viscosity μ.

The boundary conditions associated with equation [4.9] (symmetry, non-slip velocity at walls) are as follows:

$$\begin{cases} r = 0 : & \dfrac{d v_z}{d r} = 0 \\ r = R : & v_z = 0 \end{cases} \qquad [4.11]$$

where $v_z(r)$ is considered as being the target variable.

The variable physical property in the pipe is the apparent viscosity which varies according to the shear stress τ, $\mu_a(\tau)$ [3]. The associated dimensionless material function, H, is defined by:

---

3 Here, we consider the dependence of apparent viscosity on shear stress, and not on the shear rate, purely to simplify subsequent mathematical developments.

$$H = \frac{\mu_a(\tau)}{\mu_a(\tau_0)} = \frac{\mu_a(\tau)}{\mu_0} \qquad \text{[4.12]}$$

where τ_0 is any reference shear stress and μ_0 is the value of apparent viscosity at this reference shear stress.

We then make equation [4.9] dimensionless by using as repeated variables R, τ_w and μ_0. By defining:

$$v^* = \frac{v_z.\mu_0}{\tau_w.R} \quad \text{and} \quad r^* = \frac{r}{R} \qquad \text{[4.13]}$$

Equation [4.9] then becomes:

$$\left[\begin{array}{l} \tau_w.r^* + (H.\mu_0) \cdot \dfrac{d\left[\dfrac{v^*.\tau_w.R}{\mu_0}\right]}{d[r^*.R]} = 0 \\[3mm] \Leftrightarrow \tau_w.r^* + (H.\mu_0) \cdot \dfrac{\tau_w.R}{\mu_0.R} \cdot \dfrac{dv^*}{dr^*} = 0 \\[3mm] \Leftrightarrow \tau_w.r^* + (H.\tau_w) \cdot \dfrac{dv^*}{dr^*} = 0 \\[3mm] \Leftrightarrow \left(\dfrac{dv^*}{dr^*}\right) = -\dfrac{r^*}{H} \end{array}\right. \qquad \text{[4.14]}$$

where v^* is the target internal measure[4] and r^* is the measure responsible for the variations of v^*.

---

4 v^* is indeed dimensionless, as shown below:

$$\left[v^*\right] = \left[\frac{v_z.\mu_0}{\tau_w.R}\right] = \frac{\left[L.T^{-1}\right].\left[M.L^{-1}.T^{-1}\right]}{\left[M.L^{-1}.T^{-2}\right].\left[L\right]} = \left[-\right]$$

When made dimensionless, the boundary conditions lead to:

$$\begin{cases} r^* = 0: & \dfrac{dv^*}{dr^*} = 0 \\ r^* = 1: & v^* = 0 \end{cases}$$

[4.15]

Equation [4.14] can be integrated and have as a solution:

$$v^* = \int_0^{v^*} dv^* = -\int_1^{r^*} \frac{r^*}{H} \cdot dr^*$$

[4.16]

Equation [4.16] shows that the process relationship between the target internal measure v^* and the internal measure r^* depends on the dimensionless material function H, and therefore on the reference abscissa τ_0 (equation [4.12]). From a mathematical point of view, this means that v^* is a "*function of functions H*", also called a *functional of H*, which is denoted by adding the index H as follows:

$$v^* = f_H(r^*)$$

[4.17]

It is also important to recognize that v^* does not depend on the dimensional material function $\mu_a(\tau)$, but only on its dimensionless formulation, H.

4.2.3. A necessary condition for the application of a process relationship to different materials: material similarity

The conclusion drawn from the previous example can be generalized to cases which do not have a known analytical solution. Indeed, even when the evolution of a target dimensionless number is not determined by integrating differential equations associated with boundary and initial conditions, but by making a blind dimensional analysis and by fitting experimental measurements with a process relationship, any process relationship involving a target dimensionless number can be seen as the solution to a set of differential equations and boundary and initial conditions. It will, therefore, also depend on the dimensionless material function H.

As a result, we can retain that for a material with a variable physical property, *the process relationship which links the target internal measure to* *the other internal measures depends on the dimensionless material function H, and therefore on the reference abscissa p_0.*

This dependence of the process relationship on the dimensionless material function (and not on its dimensional formulation s) is one of the pillars on which the definition of the similarity between two materials will be based.

The subsequent definition of *material similarity* is:

Two materials are similar if they can be represented by one single dimensionless material function H, regardless of the concordance of their dimensional material functions.

Such a definition has an immediate practical consequence. When experiments are carried out on laboratory-scale equipment at an ambient temperature ($1/F_e$ scale) with a view to size a thermal process at high temperatures ($1/1$ scale), the search for suitable model materials is a central problem, which is frequently difficult to overcome. The fact that the model material does not need to exhibit the same dimensional material function but only meet the same dimensionless material function as the material used in the thermal process on a $1/1$ scale makes the problem easier.

The concept of material similarity and its consequences with respect to the theory of similarity will be addressed in Chapter 5. However, we should immediately draw the user's attention to the following point: the identity of an "operating point" on two scales ($1/F_e$ scale and $1/1$ scale) means conserving all of the internal measures associated with the variability of the material's physical property (to be denoted by $\{\pi_m\}$) and all of the other internal measures $\{\pi_i\}$ responsible for the causes influencing the target internal measure. It is, therefore, important:

– to choose the model material so as to conserve all of the internal measures associated with the variability of the material's physical property $\{\pi_m\}$;

– to correctly identify how the properties of the model material studied influences the target variable.

We will refer back to this point in Chapter 5.

4.3. Dimensionless material functions: standard non-dimensionalization method and invariance properties

4.3.1. *Influence of the reference abscissa and the non-dimensionalization method on the form of the dimensionless material function*

The objective here is to illustrate the fact that the graphic representation of the dimensionless material function depends, in addition to the mathematical definition of the dimensional material function, on the reference abscissa and the definition of the argument.

4.3.1.1. *Influence of the reference abscissa*

We will show the influence of the reference abscissa using the example of a Bingham fluid.

EXAMPLE 4.2.– Bingham fluid (I)

For a Bingham fluid, the dimensional material function describing the variation of apparent viscosity μ_a with the shear rate $\dot{\gamma}$ is expressed as:

$$\mu_a = \frac{\tau_y}{\dot{\gamma}} + \mu_p \qquad [4.18]$$

where τ_y is the yield stress and μ_p is the plastic viscosity of the fluid when $\tau > \tau_y$.

We will now express its dimensionless material function, H, by defining any reference shear rate $\dot{\gamma}_0$ as:

$$H = \frac{\mu_a}{\mu_0} \text{ with } \mu_0 = \mu_a(\dot{\gamma}_0) = \frac{\tau_y}{\dot{\gamma}_0} + \mu_p \qquad [4.19]$$

gives:

$$H = \frac{\mu_a}{\mu_0} = \frac{\dfrac{\tau_y}{\dot{\gamma}} + \mu_p}{\mu_0} = \frac{\tau_y}{\mu_0 \cdot \dot{\gamma}} + \frac{\mu_p}{\mu_0} \qquad [4.20]$$

We can define the dimensionless number Bi where[5]:

$$Bi = \frac{\tau_y}{\mu_0 \cdot \dot{\gamma}_0}$$ [4.21]

Combining equations [4.20] and [4.21] gives:

$$H = Bi \cdot \frac{\dot{\gamma}_0}{\dot{\gamma}} + \frac{\mu_0 - \frac{\tau_y}{\dot{\gamma}_0}}{\mu_0} = Bi \cdot \frac{\dot{\gamma}_0}{\dot{\gamma}} + 1 - \frac{\tau_y}{\dot{\gamma}_0 \cdot \mu_0}$$ [4.22]

$$= Bi \cdot \left(\frac{\dot{\gamma}_0}{\dot{\gamma}} \right) + (1 - Bi) = Bi \cdot \left(\frac{\dot{\gamma}_0}{\dot{\gamma}} - 1 \right) + 1$$

Among the various choices possible, we have chosen to express the argument v of the dimensionless material function H as follows:

$$v = \frac{\dot{\gamma}_0}{\dot{\gamma}}$$ [4.23]

The dimensionless material function associated with a Bingham fluid can, therefore, be expressed as:

$$v = \frac{\dot{\gamma}_0}{\dot{\gamma}} \quad \mapsto \quad H(v) = \frac{\mu_a}{\mu_0} = Bi \cdot v + (1 - Bi)$$ [4.24]

In Figure 4.1(a), the dimensional material function of a Bingham fluid having yield stress of $\tau_y = 40$ Pa and a plastic viscosity of $\mu_p = 20$ Pa.s has been plotted as a function of the shear rate. On this curve are also reported the points A and B of which respective coordinates, (10 s^{-1}; 24 Pa.s) and (1 s^{-1}; 60 Pa.s), are calculated, from equation [4.18], considering two reference abscissas of $\dot{\gamma}_0 = 10$ s^{-1} and $\dot{\gamma}_0 = 1$ s^{-1}, respectively. In Figure 4.1(b), the dimensionless material functions H associated with these two reference

---

5 In rheology, this dimensionless number is called the Bingham number and is traditionally noted as Bi. It has nothing to do with mass and thermal Biot numbers (as defined in Appendix 2), even if these numbers are noted as Bi_M and Bi_T respectively, or sometimes just Bi.

points are represented according to their argument v (defined in equation [4.23]).

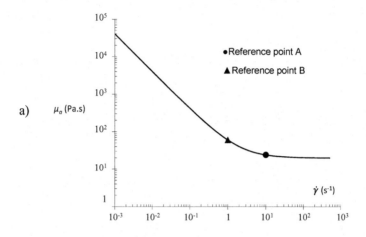

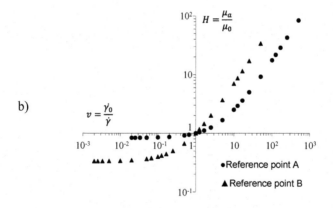

Figure 4.1. *Graphical representation of a Bingham fluid having with a yield stress of τ_y = 40 Pa and a plastic viscosity of μ_p = 20 Pa.s: a) dimensional material function; b) dimensionless material function H(v) as defined in equation [4.24]. Reference points A and B have abscissas of $\dot{\gamma}_0$ = 10 s^{-1} and $\dot{\gamma}_0$ = 1 s^{-1}, respectively*

Figure 4.1(b) clearly shows that the form of graphical representation of the dimensionless material function H associated with a Bingham fluid depends on the values of the reference shear rate (here $\dot{\gamma}_0 = 10 \text{ s}^{-1}$ for point A and $\dot{\gamma}_0 = 1 \text{ s}^{-1}$ for point B).

The conclusion drawn from this example can be generalized for all cases, namely:

For a given dimensional material function s(p) and a given definition of an argument v, the associated dimensionless material function H depends on the reference abscissa p_0 chosen. In other words, depending on the choice of p_0, the form of the graphical representations of the dimensionless material function $H(v)$ will be different.

4.3.1.2. Influence of the definition of the argument

We will reuse the previous example (Bingham fluid) to highlight this time the influence of the non-dimensionalization method.

EXAMPLE 4.2.– Bingham fluid (II)

In equation [4.22], the non-dimensionalization method chosen led to an expression of the dimensionless material function H producing a dimensionless number, Bi, and an argument v as defined in equation [4.23]. However, this dimensionless material function H can take an infinite number of other forms depending on the definition of the argument v. For example, another expression of H is:

$$H = \frac{\mu_a}{\mu_0} = 1 + \frac{v}{1 - \dfrac{v}{Bi}} \qquad [4.25]$$

where Bi is defined in the same way as in equation [4.21] and, in this case, the argument v is expressed as follows[6]:

$$v = \left(\frac{\dot{\gamma}_0 - \dot{\gamma}}{\dot{\gamma}_0} \right) Bi \qquad [4.26]$$

---

6 Later in this chapter, we will see that this argument is the one obtained when applying the standard non-dimensionalization method to the dimensional material function associated with a Bingham fluid.

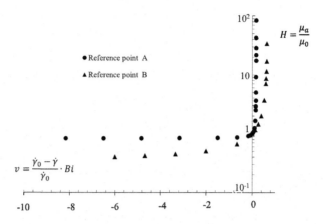

Figure 4.2. *Dimensionless material function H (defined according to equation [4.25]) as a function of its argument v (defined in equation [4.26]) for a Bingham fluid having with a yield stress of τ_y = 40 Pa and a plastic viscosity of μ_p = 20 Pa.s. Reference points A and B have abscissas of $\dot{\gamma}_0$ = 10 s^{-1} and $\dot{\gamma}_0$ = 1 s^{-1}, respectively*

The graphical representation of the dimensionless material function H in terms of its argument v as defined in equation [4.26] is given in Figure 4.2. It is clear that its form is completely different from that defined by equation [4.22] (Figure 4.1(b)).

The conclusion drawn from this example can be generalized for all cases, namely:

For a given dimensional material function s(p) and a given reference abscissa p_0, the way to define the argument v (i.e. the method of non-dimensionalization chosen) affects the form of the graphical representation of the dimensionless material function H(v).

4.3.1.3. Conclusion

In conclusion, it should be noted that the graphical representation of the dimensionless material function $H(v)$ depends on:

– the mathematical equation defining the dimensional material function $s(p)$;

– the reference abscissa p_0;

– the non-dimensionalization method (definition of the argument v).

Therefore, *it cannot be expected that the process relationship which links the target internal measure to the other internal measures is independent of the choice of the reference abscissa and the definition of the argument v (see section 4.2.2) for a physical phenomenon involving a material function s(p).*

4.3.2. Non-dimensionalization method and theory of similarity

In this section, we will answer the following question: *are there non-dimensionalization methods for the material function which do not comply with the principles of the theory of similarity?* We predict the response: *several methods are possible, but not all comply with the principles of the theory of similarity.*

As will be shown in Chapter 5, the *theory of similarity* states that only the identity of the "operating points" on both scales (for example, on a 1/1 scale for an industrial process and on a $1/F_e$ scale for laboratory experiments) is required to ensure the equality of target internal measures.

For materials with constant physical properties, this identity is established simply by maintaining the numerical values of all the internal measures which describe the configuration of the system.

For materials with variable properties, maintaining only the numerical values of these internal measures will not necessarily induce the identity of the phenomena, since the process relationship is also a functional of the dimensionless material function $H(v)$.

It will be, therefore, necessary to ensure the identity of this functional in order to guarantee the identity of the physical phenomena. However, for some cases, the identity of this functional requires some supplementary relationships to be satisfied by dimensional physical quantities. In this case, the dimensional physical quantities could not be freely selected since additional constraints between *dimensional* physical quantities are involved. The existence or non-existence of these constraints is linked to the non-dimensionalization method of the material function. So, in order to fulfill the principles of the theory of similarity, precaution should be taken with regard to the non-dimensionalization method employed.

In order to illustrate this point, we will go back to the demonstration provided by Pawlowski [PAW 91]. He used the example of a process involving the diffusion of water in a solid body when the diffusion coefficient depends on the moisture of the solid. We have made some modifications to this example, particularly by presenting it from the point of view of a drying process, since this operation is widely used in industrial agro-food processes. In addition to its practical interest, the advantage of this problem is that it can be completely formulated mathematically and numerically resolved when the geometry of the solid body is simple. The aim here is not to discuss the physical phenomena involved, but to illustrate the fact that certain non-dimensionalization methods do not guarantee that the framework necessary for applying dimensional analysis is ensured.

Let us consider as a homogeneous solid body a flat solid plate with a thickness of $2L$ whose width and length are much greater than the thickness (Figure 4.3). During the drying process (for example, by an air current), the water contained in the solid will be able to diffuse toward the surface in the direction x. The process is considered isothermal.

The diffusion coefficient D of water in this body is not constant: it depends on moisture X, defined as the mass of water contained in the solid per unit of dry solid mass. The dimensional material function used in this problem is, therefore: $D(X)$.

When $t = 0^-$ (before the process starts), the plate has a uniform moisture X_i.

When $t = 0^+$, the solid is abruptly put in contact with the air current with relative moisture ε[7] (Figure 4.3). It is assumed that the boundary conditions[8] are such that equilibrium is immediately achieved at $t = 0^+$ at the surface of the solid $(X = X_s)$.

This drying process is described by the following differential equation:

$$\frac{\partial X}{\partial t} = \operatorname{div}\left\{ D(X).\overrightarrow{\operatorname{grad}} X \right\} \qquad [4.27]$$

In the present case where the process of water diffusion is assumed to take place only in the x-direction, equation [4.27] reduces to:

---

7 Relative moisture is expressed as the ratio of partial water vapor pressure contained in the air and the vapor pressure at the same temperature.
8 By "boundary conditions", we are referring to the hydrodynamic conditions present at the surface of the solid (air convection).

$$\frac{\partial X}{\partial t} = D(X).\frac{\partial^2 X}{\partial x^2} + \frac{\partial D(X)}{\partial x}.\frac{\partial X}{\partial x}$$

[4.28]

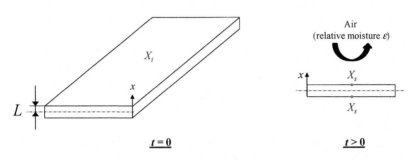

Figure 4.3. *Drying process of a flat solid plate*

The associated initial and boundary conditions are:

$$\begin{cases} t = 0, \ \forall x \qquad : X = X_i \\ \forall t > 0, \ x = L : X = X_s \end{cases}$$

[4.29]

The target variable is the moisture in the solid which varies over time t and over the thickness of the plate (described by position x): $X(x,t)$. $X(x,t)$ also depends on the dimensional material function $D(X)$.

We then introduce the following dimensionless numbers[9]:

$$\begin{cases} x^* = \dfrac{x}{L} \\ Fo = \dfrac{t.D(X_s)}{L^2} \end{cases}$$

[4.30]

Let us take the reduced moisture X^* as a target internal measure:

$$X^* = \frac{X - X_s}{X_i - X_s}$$

[4.31]

---

9 It should be noted that Fo refers to the mass Fourier number (see Appendix 2).

The dimensionless material function H is defined by considering as reference moisture the surface moisture X_s, giving:

$$H(X^*) = \frac{D(X)}{D(X_s)} \qquad [4.32]$$

The combination of equations [4.28]–[4.29] and [4.30]–[4.32] leads to the dimensionless formulation[10]:

$$\frac{\partial X^*}{\partial Fo} = \frac{\partial}{\partial x^*}\left\{ H(X^*).\overrightarrow{\mathrm{grad}}X^* \right\} \qquad [4.33]$$

which can also be written in this case:

$$\frac{\partial X}{\partial Fo} = H(X^*).\frac{\partial^2 X^*}{\partial x^{*2}} + \frac{\partial H(X^*)}{\partial x^*}.\frac{\partial X^*}{\partial x^*} \qquad [4.34]$$

With the following initial and boundary conditions:

$$\begin{cases} Fo = 0, & \forall x^* & : X^* = 1 \\ \forall Fo > 0, & x^* = 1 & : X^* = 0 \end{cases} \qquad [4.35]$$

The resolution of equations [4.34]–[4.35] provides the target internal measure X^*, which represents the time variation of the moisture field in the solid plate. Crank [CRA 75] has described the solution of equation [4.34] for different mathematical forms of the dimensional material function $D(X)$.

It is important to note that the target internal measure, X^*, depends on the internal measures Fo and x^*, and also on the dimensionless material function

---

10 It is interesting to note that an identical expression would be obtained if the problem was formulated in terms of heat transfer (e.g. the cooling or heating of a flat plate), instead of mass transfer. Moisture would then be replaced by temperature and the diffusion coefficient by thermal diffusivity. If thermal diffusivity is constant, there is a graphic solution to this problem: the famous Gurney-Lurie charts [GUR 23] in which the thermal Fourier number and a reduced temperature T^* form the axes of the abscissas and ordinates of the chart, respectively. Conversely, these charts are perfectly suitable for mass transfer provided that the diffusion coefficient is constant.

H. As in the example of the velocity profile in a pipe (section 4.2.2), the target internal measure X^* is a *functional* of *H*:

$$X^* = f_H(x^*, Fo) \tag{4.36}$$

For a dimensional analysis carried out for materials with constant physical properties, the evolution of X^* depends solely on the collection of specific values taken by Fo and x^*.

For a dimensional analysis carried out for materials with variable physical properties, relationship [4.36] indicates that the evolution of X^* depends on the collection of particular values taken by Fo and x^*, as well as on the dimensionless material function *H*. The *identity of the target internal measure X**, therefore, assumes that the values of *H* and its derivative *H'* remain the same whatever the values of X_i and X_s. We will demonstrate (*reductio ad absurdum*) that this is not the case. To do so, we will study the behavior of the dimensionless material function *H* in the vicinity of the reference abscissa X_s ($X^* = 0$), namely its expression and that of its derivative at this point.

According to equation [4.32], we have:

$$H(X^* = 0) = \frac{D(X_s)}{D(X_s)} = 1 \tag{4.37}$$

Therefore, at the reference abscissa X_s ($X^* = 0$), the value taken by the dimensionless material function *H* is then always constant (it is equal to 1).

However, this is not the case for its derivative *H'* (the slope at the origin of the function *H*). Indeed, using equation [4.31], it is possible to express *X* in terms of X^*:

$$X = X^* \cdot (X_i - X_s) + X_s \tag{4.38}$$

From equation [4.32], it can be deduced that:

$$H(X^*) = \frac{D(X)}{D(X_s)} = \frac{D(X^* \cdot (X_i - X_s) + X_s)}{D(X_s)} \tag{4.39}$$

hence

$$H'(X^*) = \frac{\partial H}{\partial X^*} = \frac{1}{D(X_s)} \cdot \frac{\partial D}{\partial X^*} \cdot (X_i - X_s) \qquad [4.40]$$

Therefore, at the reference abscissa:

$$H'(X^* = 0) = \left(\frac{\partial H}{\partial X^*}\right)_{X^*=0} = \frac{1}{D(X_s)} \cdot \left(\frac{\partial D}{\partial X^*}\right)_{X=X_s} \cdot (X_i - X_s) \qquad [4.41]$$

The derivative of the dimensionless material function, H', does not therefore take a unique value, since it depends on the values of the dimensional physical quantities X_i and X_s. This means that the graphical representation of the dimensionless material function $H(X^*)$ is not unique as it is influenced by the initial moisture values and those at the surface of the solid, X_i and X_s.

This demonstration, carried out in the vicinity of the argument $X^* = 0$, shows that, depending on the values of X_i and X_s, the target internal measure (X^*) is not necessarily unique for a given "operating point" (namely, for given values of X^* and Fo)[11]. These supplementary constraints to be fulfilled by two independent dimensional physical quantities do not comply with the principles of the theory of similarity as given for materials with constant physical properties.

Thus, we will retain that *all the methods for the non-dimensionalization of the material function do not necessarily lead to compliance with the principles of the theory of similarity.*

4.3.3. Standard non-dimensionalization method

Using the work of Pawlowski [PAW 91] as a basis, we will now demonstrate how the non-dimensionalization of the material function must be

---

11 Attentive users will have noticed that, in this example of a drying process of a solid body, the target internal measure (X^*) is also the argument v of the dimensionless material function H. Pawlowski [PAW 91] used a very specific case, but rest assured, that does not jeopardize the illustrative nature of this example.

carried out so that it complies with the principles of the theory of similarity. This consists of applying a specific non-dimensionalization method, which we will call the *standard non-dimensionalization method*.

4.3.3.1. Definition of the standard dimensionless material function

Let the dimensional material function be $p \mapsto s(p)$ which describes the variation of the physical property s with variable p.

Let p_0 be the reference abscissa and $(p_0 ; s(p_0))$ the reference point, so that $s(p_0) \neq 0$ and $\left(\dfrac{ds}{dp} \right)_{p=p_0} \neq 0$.

Whatever the material function $s(p)$, its standard non-dimensionalization method consists of defining the *argument* of the dimensionless material function whereby:

$$u = a_0 .(p - p_0) \tag{4.42}$$

It is denoted as u in order to distinguish it from an argument v defined by a method other than standard non-dimensionalization method.

a_0 is called the coefficient of physical property s relative to variable p and is expressed as follows:

$$a_0 = \frac{1}{s(p_0)} . \left(\frac{ds(p)}{dp} \right)_{p=p_0} \tag{4.43}$$

a_0 is the slope of the dimensional material function $s(p)$ calculated in $p = p_0$ and divided by $s(p_0)$[12].

---

12 In terms of notation, we differentiate ourselves here from Pawlowski [PAW 91] who defined this coefficient a_0 as the inverse of that described by equation [4.43]. Indeed, we believe that defining the coefficient a_0 as we have done gives it an explicit and more direct physical meaning (see equation [4.44]).

It is interesting to note that in the case of the dependence of a physical property (e.g. the Newtonian viscosity μ of a fluid) on temperature θ, this coefficient is traditionally called the temperature coefficient and denoted as γ so that:

$$\gamma = \frac{1}{\mu(\theta)} \cdot \left(\frac{d\mu}{d\theta} \right)_{\theta} \qquad [4.44]$$

We will now denote by w the *standard dimensionless material function* in order to differentiate it from the dimensionless material function H defined using a different method. For $s(p_0) \neq 0$ and $\left(\dfrac{d s}{d p} \right)_{p=p_0} \neq 0$, we have

$$H(v) = \frac{s(p)}{s(p_0)} = w(u) \qquad [4.45]$$

If its argument is defined as

$$v = u = a_0 \cdot (p - p_0) = \frac{1}{s(p_0)} \cdot \left(\frac{d s(p)}{d p} \right)_{p=p_0} \cdot (p - p_0) \qquad [4.46]$$

This non-dimensionalization method ensures that *in the vicinity of the reference abscissa (u = 0), the dimensionless material function w(u) and its derivative w'(u) are constants equal to 1 (whatever the form of its function s(p)), that is to say that they are independent of the dimensional physical quantities which influence the system.* Such a method enables us to overcome the problem of non-compliance with the principles of the theory of similarity. We will illustrate this in section 4.3.3.2.

It should be noted that equations [4.45] and [4.46] can be grouped into a single equation:

$$w(u) = \frac{s\left[p_0 + \dfrac{u}{a_0} \right]}{s(p_0)} = \frac{s\left[p_0 + u \cdot \dfrac{s(p_0)}{\left(\dfrac{d s(p)}{d p} \right)_{p=p_0}} \right]}{s(p_0)} \qquad [4.47]$$

In order to illustrate the concept of standard dimensionless material function, we will re-examine the example of the pseudoplastic fluid.

EXAMPLE 4.1.– Pseudoplastic fluid (II)

We have seen that the dimensional material function of a pseudoplastic fluid is given by equation [4.4]. It is graphically represented in Figure 4.4(a) for a pseudoplastic fluid with a consistency of $k = 0.1$ Pa.s^n and a flow index of $n = 0.4$.

We previously established the expression of the dimensionless material function in relation to the shear rate (equation [4.6]) using its definition (equation [4.4]). This is repeated below, but this time it is denoted as w since we will associate it with an argument u defined according to the standard non-dimensionalization method:

$$w(u) = \frac{\mu_a}{\mu_0} = \frac{k \cdot \dot{\gamma}^{n-1}}{k \cdot \dot{\gamma}_0^{n-1}} = \left(\frac{\dot{\gamma}}{\dot{\gamma}_0}\right)^{n-1} \qquad [4.48]$$

where μ_0 is the value of the apparent viscosity at the reference shear rate $\dot{\gamma}_0$ ($\dot{\gamma}_0 \neq 0$).

The application of this method requires the calculation of its argument u using equation [4.46], whereby:

$$u = a_0 \cdot (\dot{\gamma} - \dot{\gamma}_0) = \frac{1}{\mu_a(\dot{\gamma}_0)} \cdot \left(\frac{d\mu_a(\dot{\gamma})}{d\dot{\gamma}}\right)_{\dot{\gamma}=\dot{\gamma}_0} \cdot (\dot{\gamma} - \dot{\gamma}_0) \qquad [4.49]$$

Given that:

$$\begin{cases} \left(\dfrac{d\mu_a(\dot{\gamma})}{d\dot{\gamma}}\right)_{\dot{\gamma}=\dot{\gamma}_0} = (n-1).k.\dot{\gamma}_0^{n-2} \\[4mm] a_0 = \dfrac{(n-1).k.\dot{\gamma}_0^{n-2}}{k.\dot{\gamma}_0^{n-1}} = (n-1).\dot{\gamma}_0^{-1} \end{cases} \qquad [4.50]$$

makes:

$$u = (n-1).\dot{\gamma}_0^{-1}.(\dot{\gamma} - \dot{\gamma}_0) = \frac{\dot{\gamma} - \dot{\gamma}_0}{\dot{\gamma}_0}.(n-1) \qquad [4.51]$$

From equation [4.51], it can be deduced that:

$$\dot{\gamma} = \dot{\gamma}_0 \cdot \left[\frac{u}{n-1} + 1 \right] \qquad [4.52]$$

Let us introduce equation [4.52] into equation [4.48]:

$$w(u) = \left(\frac{\dot{\gamma}}{\dot{\gamma}_0} \right)^{n-1} = \left(\frac{\dot{\gamma}_0 \cdot \left[\frac{u}{n-1} + 1 \right]}{\dot{\gamma}_0} \right)^{n-1}$$

$$= \left(\frac{u}{n-1} + 1 \right)^{n-1} \qquad [4.53]$$

Therefore, the *standard* dimensionless material function w associated with a pseudoplastic fluid is expressed as follows:

$$u = \frac{\dot{\gamma} - \dot{\gamma}_0}{\dot{\gamma}_0}.(n-1) \quad \mapsto \quad w(u) = \frac{\mu_a}{\mu_0} = \left(\frac{u}{n-1} + 1 \right)^{n-1} \qquad [4.54]$$

Figure 4.4(b) shows the graphical representation of $w(u)$ for a pseudoplastic fluid with a consistency of $k = 0.1$ Pa.s^n and a flow index of $n = 0.4$. It is clear that the conditions of standardization have been fulfilled, given that $u = 0$:

– the standard dimensionless material function w is constant, equal to 1;

– its derivative $\dfrac{\mathrm{d}\,w}{\mathrm{d}\,u}$ which is the slope of the curve representing $w(u)$ (indicated in Figure 4.4(b) by a double arrow), is also constant, equal to 1.

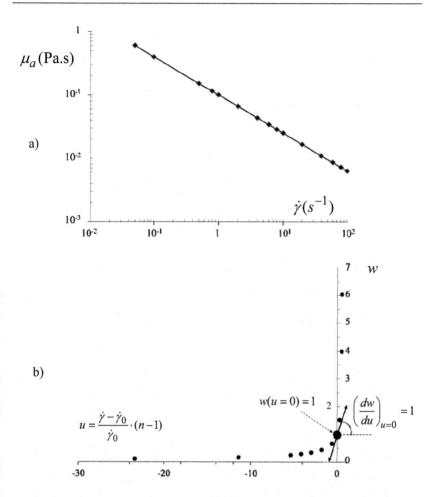

Figure 4.4. *Pseudoplastic fluid having a consistency of k = 0.1 Pa.s^n and a flow index of n = 0.4: a) dimensional material function and b) standard dimensionless material function w(u) defined in equation [4.54] (the reference shear rate chosen is $\dot{\gamma}_0$ =1 s^{-1})*

4.3.3.2. *Illustration of the interest of the standard dimensionless material function with respect to the principles of the theory of similarity*

Let us refer back to the drying process of a flat solid plate (Figure 4.3) where the dimensional material function is $D(X)$ and the target variable is the moisture of the solid X.

We will take any reference abscissa X_0 so that $D(X_0) \neq 0$ and $D'(X_0) \neq 0$.

The application of the standard non-dimensionalization method (equation [4.46]) leads to a definition of the associated standard dimensionless material function w as:

$$u \quad \mapsto \quad w(u) = \frac{D(X)}{D(X_0)} \qquad [4.55]$$

and its argument u as:

$$u = a_0.(X - X_0) = \frac{X - X_0}{D(X_0)}.D'(X_0) \qquad [4.56]$$

where $D'(X_0)$ is the derivative of the dimensional material function in relation to X in X_0:

$$D'(X_0) = \left(\frac{dD(X)}{dX} \right)_{X=X_0} \qquad [4.57]$$

u is the target internal measure[13].

We can also write:

$$\begin{cases} x^* = \dfrac{x}{L} \\ Fo = \dfrac{t.D(X_0)}{L^2} \end{cases} \qquad [4.58]$$

---

13 Like the reduced moisture X^* in section 4.3.2, the argument u of the standard dimensionless material function w is here also the target internal measure.

The combination of equations [4.28]–[4.29] and [4.55]–[4.57] leads to the following dimensionless formulation:

$$\frac{\partial u}{\partial Fo} = \frac{\partial}{\partial x^*}\left\{ w(u).\overrightarrow{\text{grad}\,u}\right\}$$ [4.59]

which in this case can also be written as:

$$\frac{\partial u}{\partial Fo} = w(u).\frac{\partial^2 u}{\partial x^{*2}} + \frac{\partial w(u)}{\partial x^*}.\frac{\partial u}{\partial x^*}$$ [4.60]

With the following initial and boundary conditions:

$$\begin{cases} Fo = 0, \quad \forall x^* \;:\; u = u_i = \dfrac{X_i - X_0}{D(X_0)}.D'(X_0) \\[3mm] \forall\, Fo > 0, \, x^* = 1 \;:\; u = u_s = \dfrac{X_s - X_0}{D(X_0)}.D'(X_0) \end{cases}$$ [4.61]

The resolution of equations [4.60]–[4.61] provides u which represents the time variation in the moisture field in the solid plate. u depends on the internal measures Fo, x^*, u_i and u_s, and is a functional of the standard dimensionless material function w:

$$u = f_w\left(x^*, Fo, u_i, u_s\right)$$ [4.62]

For the reasons given previously, the identity of this solution requires the values of w and its derivative, $\dfrac{d\,w}{d\,u}$, to remain constant whatever the values of X_0, X_i and X_s.

Let us determine the standard dimensionless material function w in the vicinity of the reference abscissa X_0 where $u = 0$.

$$w(X^* = 0) = \frac{D(X = X_0)}{D(X_0)} = 1$$ [4.63]

Equation [4.63] shows that w is constant (equal to 1) when $X = X_0$.

Let us now determine the derivative of the standard dimensionless material function $\dfrac{\mathrm{d}\,w}{\mathrm{d}\,u}$ in the vicinity of the reference abscissa X_0.

According to equation [4.56], this gives:

$$X = X_0 + u.\frac{D(X_0)}{D'(X_0)} \qquad\qquad [4.64]$$

If $D(X_0) \neq 0$ and $D'(X_0) \neq 0$, it can be deduced that:

$$
\left\{
\begin{aligned}
(w')_{u=0} &= \left(\frac{\partial w}{\partial u}\right)_{u=0} = \left(\frac{\partial}{\partial u}\left(\frac{D(X\ )}{D(X_0)}\right)\right)_{u=0} \\[2mm]
(w')_{u=0} &= \left(\frac{\partial}{\partial u}\left(\frac{D\!\left(X_0 + u.\dfrac{D(X\ )}{D'(X_0)}\right)}{D(X_0)}\right)\right)_{u=0} \\[2mm]
(w')_{u=0} &= \frac{D(X\ )}{D'(X_0)}.\left(\frac{\partial D}{\partial u}\right)_{u=0}.\frac{1}{D(X_0)} = 1
\end{aligned}
\right.
\qquad [4.65]
$$

Therefore, just like w, w' is constant (equal to 1) when $X = X_0$. This example helps illustrate the fact that in the vicinity of the reference abscissa ($X = X_0$ or $u = 0$), the standard dimensionless material function and its derivative are always constant and independent of the values of the dimensional quantities.

By using the standard method to make the material function dimensionless, it is thus ensured that in the vicinity of the reference abscissa:

– the evolution of X^* is solely dependent on the collection of the specific values taken $Fo,\ X^*$, u_i and u_s;

– therefore, the process relationship becomes independent of H.

In this way, the problem of non-compliance with the principles of the theory of similarity can be overcome.

4.3.3.3. *Origin of the standard dimensionless material function*

The dimensional material function $s(p)$ can be expressed in the form of a Taylor series:

$$p \mapsto s(p) = C_0 + C_1 \cdot (p - p_0) + C_2 \cdot (p - p_0)^2 + \sum_{k=3}^{+\infty} C_k \cdot (p - p_0)^k \qquad [4.66]$$

where p_0 is a reference abscissa so that $s(p_0) \neq 0$ and $\left(\dfrac{d\,s}{d\,p} \right)_{p=p_0} \neq 0$.

The coefficients C_k are defined by:

$$\begin{cases} k = 0 \quad : \quad C_0 = s(p_0) \\ k \geq 1 \quad : \quad C_k = \dfrac{1}{k!} \left(\dfrac{\partial^k s}{\partial p^k} \right)_{p=p_0} \end{cases} \qquad [4.67]$$

where $k!$ represents the factorial of the natural integer k. It is also possible to write:

$$\begin{cases} w(u) = \dfrac{s(p)}{s(p_0)} \\ w(u) = 1 + \dfrac{(p - p_0)}{1! s(p_0)} \left(\dfrac{\partial s}{\partial p} \right)_{p=p_0} + \dfrac{(p - p_0)^2}{2! s(p_0)} \left(\dfrac{\partial^2 s}{\partial p^2} \right)_{p=p_0} \\ \quad + \dfrac{(p - p_0)^3}{3! s(p_0)} \left(\dfrac{\partial^3 s}{\partial p^3} \right)_{p=p_0} + \sum_{k=4}^{+\infty} \dfrac{(p - p_0)^k}{k! s(p_0)} \left(\dfrac{\partial^k s}{\partial p^k} \right)_{p=p_0} \end{cases} \qquad [4.68]$$

Given that in the standard non-dimensionalization method (equation [4.46]):

$$u = \frac{(p - p_0)}{s(p_0)} \cdot \left(\frac{ds}{dp} \right)_{p=p_0} \qquad [4.69]$$

Equation [4.68] becomes:

$$
\begin{cases}
w(u) = \dfrac{s(p)}{s(p_0)} \\[2em]
w(u) = 1 + u + u^2 \cdot \dfrac{[s(p_0)]^{2-1} \cdot \left(\dfrac{\partial^2 s}{\partial p^2}\right)_{p=p_0}}{2! \left[\left(\dfrac{\partial s}{\partial p}\right)_{p=p_0}\right]^2} + u^3 \cdot \dfrac{[s(p_0)]^{3-1} \cdot \left(\dfrac{\partial^3 s}{\partial p^3}\right)_{p=p_0}}{3! \left[\left(\dfrac{\partial s}{\partial p}\right)_{p=p_0}\right]^3} \\[2em]
\quad + \displaystyle\sum_{k=4}^{+\infty} u^k \cdot \dfrac{[s(p_0)]^{k-1} \cdot \left(\dfrac{\partial^k s}{\partial p^k}\right)_{p=p_0}}{k! \left[\left(\dfrac{\partial s}{\partial p}\right)_{p=p_0}\right]^k}
\end{cases}
\qquad [4.70]
$$

We can define the function q_k for $k \geq 1$ as:

$$
q_k = \dfrac{[s(p_0)]^{k-1} \cdot \left(\dfrac{\partial^k s}{\partial p^k}\right)_{p=p_0}}{k! \left[\left(\dfrac{\partial s}{\partial p}\right)_{p=p_0}\right]^k}
\qquad [4.71]
$$

It can then be shown that, if $s(p_0) \neq 0$ and $\left(\dfrac{d s}{d p}\right)_{p=p_0} \neq 0$, a standard dimensionless material function can always be expressed as:

$$
w(u) = \dfrac{s(p)}{s(p_0)} = 1 + u + u^2 . q_2 + u^3 . q_3 + \sum_{k=4}^{+\infty}\left[u^k . q_k\right]
\qquad [4.72]
$$

Within the first order and in the vicinity of $u = 0$, whatever the form of the dimensional material function, the standard dimensionless material function can, therefore, be approximated by:

$$w(u) = \frac{s(p)}{s(p_0)} = 1 + u \quad \text{if} \quad s(p_0) \neq 0 \text{ and } \left(\frac{ds}{dp}\right)_{p=p_0} \neq 0, \quad\quad [4.73]$$

This shows that the values of the standard dimensionless material function and its derivative are always constant (equal to 1) in the vicinity of the reference abscissa:

$$w(0) = \left(\frac{dw}{du}\right)_{u=0} = 1 \quad\quad [4.74]$$

COMMENT.–

If the coefficients C_0 or C_1 are equal to zero in the Taylor series (if $s(p_0)=0$ and $\left(\dfrac{ds}{dp}\right)_{p=p_0} = 0$), we are faced with a "degenerate" case. Consequently, equations [4.66] and [4.67] must be initially transformed whereby:

If $C_0=0$

$$\frac{s(p) - s(p_0)}{(p - p_0)} = C_1 + C_2 \cdot (p - p_0) + \sum_{k=3}^{+\infty} C_k \cdot (p - p_0)^k \quad\quad [4.75]$$

Moreover, if $C_1=0$,

$$\frac{s(p) - s(p_0)}{(p - p_0)^2} = C_2 + C_3 \cdot (p - p_0) + \sum_{k=4}^{+\infty} C_k \cdot (p - p_0)^k \quad\quad [4.76]$$

The left-hand side of equation [4.75] or equation [4.76] which will then be subjected to the standard non-dimensionalization method, as with the material function beforehand $s(p)$. This can be generalized so that:

– transforming the Taylor series in such a way that the first two successive coefficients are not zero;

– the resultant function of this series is thus the standard representation of the left-hand term.

4.3.3.4. *Conclusion*

The standard non-dimensionalization method is based on an approximation of the dimensional material function $s(p)$ in the vicinity of the reference abscissa p_0 (by a Taylor series), normalized by $s(p_0)$. As a result, *$w(u)$ only has a meaning in the vicinity of $u = 0$.*

The standard non-dimensionalization method has two advantages:

– first, it provides the *theoretical framework in order to comply with the principles of the theory of similarity* in the case of materials with variable physical properties;

– second, *it defines a repeatable and applicable procedure whatever the dimensional material function.*

Finally, it is important to remember, as defined in equation [4.46], *that the curve which represents the standard dimensionless material function remains dependent on the chosen reference abscissa.*

We will illustrate this final point (i.e. the dependence of the standard dimensionless material function on the chosen reference abscissa) by using an example: the variation in the density of water with temperature.

EXAMPLE 4.3.– Water density according to temperature

The density of water ρ depends on the temperature at a given pressure. We can, therefore, define the dimensional material function $\rho(\theta)$.

The variation in water density ρ (at atmospheric pressure) is represented in Figure 4.5(a).

Two reference abscissas θ_0 have been randomly chosen: $\theta_0 = 293.15$ K (20°C) and $\theta_0 = 333.15$ K (60°C). Points A and B are, respectively, associated with these abscissas.

The standard dimensionless material functions associated with these two reference points are represented in Figure 4.5(b)[14]. It is interesting to note that:

---

14 These standard dimensionless material functions were obtained following an identical procedure to the one described in the examples in section 4.4.2.2. In these examples, like in the case of the dependence of water density on temperature, there is no known analytical expression of the material function.

– both curves cross at point $u = 0$;

– in accordance with the standard non-dimensionalization method (equation [4.46]), in $u = 0$, both curves cut the axes of the ordinates at $w = 1$ and have a gradient of 1;

– however, as soon as we move away from $u = 0$, the curves describing the standard dimensionless material function no longer coincide, according to the reference abscissa chosen.

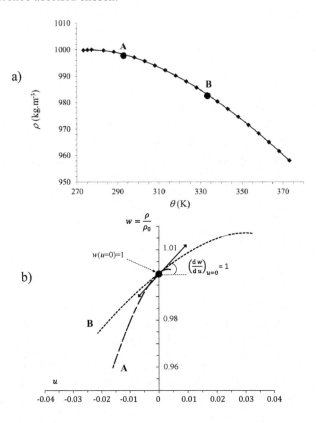

Figure 4.5. *Graphical representation of the variation of water density at atmospheric pressure with temperature a) dimensional material function. b) Standard dimensionless material function. Reference points A and B have abscissas of $\theta_0 = 293.15$ K and $\theta_0 = 333.15$ K, respectively*

4.3.4. *Standard material functions with specific properties: reference-invariant standard material functions*

In this section, we will show that:

– there are some classes of standard dimensionless material functions, described by very specific mathematical equations, which do not depend on the reference abscissa;

– it is also possible to identify them using the form of the associated dimensional material functions.

The interest for dimensional analysis of these specific material functions will be discussed in section 4.4.

4.3.4.1. *Form of the dimensionless material functions leading to invariance properties*

The existence of standard dimensionless material functions which do not depend on the reference abscissa can be proven mathematically. This uses transitivity properties of standard dimensional material functions. These mathematical demonstrations were developed by Pawlowski [PAW 91] and are provided in Appendix 3.

Therefore, it is possible to demonstrate that *there are standard dimensionless material functions which do not depend on the chosen reference abscissa.*

These are called *invariant standard dimensionless material functions* and will be denoted by ϕ, in order to distinguish it from non-invariant standard dimensionless material functions denoted by *w*.

It can be mathematically proven that *only two families* of standard dimensionless material functions have these invariant properties, which are described by the following equations:

$$\phi(u) = \left(1 + \beta.u\right)^{1/\beta}$$ [4.77]

where β is a constant so that $\beta \neq 0$

and $\phi(u) = \exp(u)$ [4.78]

Therefore, *whatever the reference abscissa chosen*, the graphical representation of the reference-invariant standard dimensionless material function $\phi(u)$ is identical.

In the following example, we will illustrate this property in a case where there is a known analytical expression of the material function: a pseudoplastic fluid. In the examples reported in section 4.4.2.2, we will show the procedure to follow when the analytical expression of the material function is not known.

EXAMPLE 4.1.– Pseudoplastic fluid (III)

We have seen (equation [4.43]) that the standard dimensionless material function associated with a pseudoplastic fluid can be expressed as:

$$w(u) = \left(\frac{u}{n-1} + 1 \right)^{n-1} \tag{4.79}$$

It can be put in the following form:

$$w(u) = \left(1 + \beta \cdot u\right)^{1/\beta} \text{ with } \beta = \frac{1}{n-1} \tag{4.80}$$

Consequently, the standard dimensionless material function associated with a pseudoplastic fluid is reference-invariant, and therefore does not depend on the reference shear rate $\dot{\gamma}_0$ chosen.

Figure 4.6 shows the standard dimensionless material function associated with a pseudoplastic fluid having a consistency of $k = 0.1$ Pa.s^n and a flow index of $n = 0.4$. Two reference points, A and B, were arbitrarily chosen, corresponding to the reference shear rates $\dot{\gamma}_0$ equal to 1 s^{-1} and 10 s^{-1}, respectively. Unlike with the case of water density (Figure 4.5(b)), it can be observed that, whatever the value of $\dot{\gamma}_0$, the graphical representation of the associated standard dimensionless material functions is identical over the entire range of u and not only in the vicinity of $u = 0$.

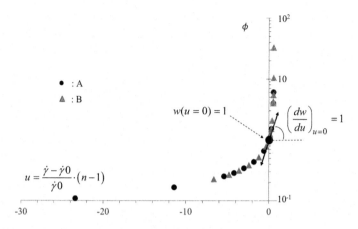

Figure 4.6. *Graphical representation of the standard dimensionless material function of a pseudoplastic fluid having a consistency of k = 0.1 Pa.s^n and a flow index of n = 0.4. Reference points A and B have abscissas of $\dot{\gamma}_0 = 1$ s^{-1} and $\dot{\gamma}_0 = 10$ s^{-1}, respectively*

4.3.4.2. Form of dimensional material functions leading to invariance properties

Let us distinguish two cases, depending on whether the reference-invariant standard dimensionless material function is described in equation [4.77] ($\beta \neq 0$) or [4.78].

1) *Invariant standard dimensionless material function described in equation [4.77] ($\beta \neq 0$)*

In this case, the reference-invariant standard dimensionless material function can be expressed by:

$$\phi(u) = \frac{s(p)}{s(p_0)} = (1 + \beta \cdot u)^{1/\beta} \qquad [4.81]$$

According to equation [4.46], the argument u is defined by:

$$u = a_0 \cdot (p - p_0) = \frac{1}{s(p_0)} \cdot \left(\frac{d\,s(p)}{d\,p} \right)_{p=p_0} \cdot (p - p_0) \qquad [4.82]$$

where a_0 is the coefficient of the physical property s relating to variable p.

Equation [4.81] thus becomes:

$$\phi(u) = \frac{s(p)}{s(p_0)} = \left[1 + \beta \cdot a_0 \cdot (p - p_0)\right]^{1/\beta} \tag{4.83}$$

After rearrangement, this gives:

$$
\begin{aligned}
s(p) &= s(p_0) \cdot \left[(1 - \beta \cdot a_0 \cdot p_0) + \beta \cdot a_0 \cdot p\right]^{1/\beta} \\
&= \left[\left([s(p_0)]^{\beta} \cdot (1 - \beta \cdot a_0 \cdot p_0)\right) + \left([s(p_0)]^{\beta} \cdot \beta \cdot a_0\right) \cdot p\right]^{1/\beta}
\end{aligned}
\tag{4.84}
$$

The form of equation [4.84] is:

$$s(p) = \left[A + B \cdot p\right]^{C} \tag{4.85}$$

where A, B and C are the constants so that

$$
\begin{cases}
A = [s(p_0)]^{\beta} \cdot (1 - \beta \cdot a_0 \cdot p_0) \\
B = [s(p_0)]^{\beta} \cdot \beta \cdot a_0 \\
C = \dfrac{1}{\beta}
\end{cases}
\tag{4.86}
$$

2) *Reference-invariant standard dimensionless material function described by equation [4.78])*

In this case, the reference-invariant standard dimensionless material function can be expressed by:

$$\phi(u) = \frac{s(p)}{s(p_0)} = \exp(u) \tag{4.87}$$

Using equation [4.82], equation [4.80] can be transformed as follows:

$$
\begin{aligned}
s(p) &= s(p_0) \cdot \exp\left(a_0 \cdot (p - p_0)\right) \\
&= \exp\left(\ln[s(p_0)] - a_0 \cdot p_0 + a_0 \cdot p\right) \\
&= \exp\left(\left[\ln[s(p_0)] - a_0 \cdot p_0\right] + [a_0] \cdot p\right)
\end{aligned}
\tag{4.88}
$$

The form of equation [4.88] is:

$$s(p) = \exp[A + B \cdot p]$$
[4.89]

where A and B are constants so that:

$$\begin{cases} A = \ln[s(p_0)] - a_0 \cdot p_0 \\ B = a_0 \end{cases}$$
[4.90]

3) *Conclusion*

Sections 4.3.4.1 and 4.3.4.2 demonstrated that if a standard dimensionless material function takes the form of equation [4.77] or [4.78], then the associated dimensional material function takes the form of equation [4.85] or [4.89]. It is possible to show that conversely if a standard dimensional material function takes the form of equation [4.85] or [4.89], then the standard dimensionless material function takes the form of equation [4.77] or [4.78].

Therefore, a *necessary* and *sufficient condition* to be satisfied so that a standard dimensionless material function *does not depend on the reference abscissa p_0 is that its dimensional material function takes the form of*

$$s(p) = (A + B.p)^C$$
[4.91]

or $$s(p) = \exp(A + B.p)$$
[4.92]

where A, B and C are the constants.

This result can be very useful in practice. Indeed, the analytical expression of the standard dimensionless material function in terms of the argument u, $w(u)$ is not always simple to obtain (this will be illustrated in the examples in the following section). As a result, *to know whether a standard dimensionless material function depends on the reference abscissa, it is often much simpler to observe the mathematical expression of the dimensional material function, rather than its dimensionless formulation.*

4.4. How to construct the π – space in the case of a process involving a material with a variable physical property

In sections 4.1 to 4.3, we saw that for a process involving a material with a variable physical property:

– the target internal measure is a functional of the dimensionless material function, and not of the dimensional material function $s(p)$;

– two materials are similar if they can be represented by a *single* and *unique* dimensionless material function, regardless of their dimensional material functions (concept of material similarity);

– the standard non-dimensionalization method is based on an approximation of the dimensional material function $s(p)$ in the vicinity of the reference abscissa p_0 (a Taylor series), normalized by $s(p_0)$. The standard dimensionless material function which results, $w(u)$, therefore only has a meaning in the vicinity of $u = 0$;

– the standard non-dimensionalization method defines:

- the theoretical framework which enables the principles of the theory of similarity to be respected in the vicinity of the reference abscissa for a material with a variable property,

- a procedure which can be repeated and applied whatever the physical property being studied,

- the standard dimensionless material function depends on the reference abscissa, except when it has invariance properties (see equations [4.77] and [4.78]).

The interest of establishing the standard dimensionless material function $w(u)$ is to offer the possibility of identifying the *configuration of the material*, which is the *set of internal measures linked to the material's variable physical property*.

This configuration will then be added to the initial configuration of the system, which corresponds to configuration which would have been established for an identical process where the physical property could be considered as constant. Therefore, the number of internal measures characterizing the causes responsible for the variation of the target internal measure will be increased if a material has a variable physical property.

The aim of this section is to give the user the tools for defining the complete set of dimensionless numbers, including the configuration of the material.

4.4.1. *Determining the dimensionless numbers and/or physical quantities to add to the initial relevance list*

This is a summary of the actions which need to be undertaken in order to unambiguously establish the complete space of dimensionless numbers (complete configuration) influencing the target internal measure in a process which involves a material with a variable physical property. This objective is accomplished by the following five-step strategy, as explained in more detail below.

To establish *the complete list of physical quantities influencing the target variable in the presence of a material with a variable physical property*, it is necessary to list:

– Step 1: all the independent physical quantities, except the variable physical property of the material s, influencing the target variable;

– Step 2: *the reference abscissa p_0* except if the material function has invariance properties;

– Step 3: $s(p_0)$, the material's variable physical property $s(p)$ calculated with a reference abscissa p_0;

– Step 4: *the set of dimensionless numbers, denoted by $\{\pi_m\}$, which appears in the expression of the dimensionless argument u of the standard dimensionless material function w except the ratio (p/p_0), or, the numerical value of the dimensionless number $\{\pi_m\}$.*

The general expression of $\{\pi_m\}$ can be noted[15]:

$$\{\pi_m\} = \{a_0 \cdot p_0\}$$

---

15 We should note that Pawlowski [PAW 91], who was the pioneer of this approach, has a different method of listing the rules in Step 4. He recommends listing a_0 at this stage. We chose the formulation $\{\pi_m\}=\{a_0.p_0\}$ which is equivalent, within the constant p_0, in order to make it more generic. This is especially the case when a mathematical law is available for the dimensionless material function and when the objective is to make explicit the set of dimensionless numbers which define the configuration of the material. This is illustrated in the final example in Chapter 5.

– Step 5: the final step consists of establishing the *new list* of relevant physical quantities to form the complete set of dimensionless numbers $\{\pi_i\}$ responsible for variation in the target internal measure, including the internal measures or the physical quantities linked to the material's variable property (material configuration).

4.4.1.1. *Step 1*

The first step deals with the establishment of the list of relevant independent physical quantities, except the material's variable physical property s, which could have an influence on the target variable. This list is the same as the one which would have been established for an identical process where the material's physical property was constant. We will name this the *"initial relevance list"*. Indeed, there is no physical reason to eliminate any of these physical quantities when the physical property of the material s depends on variable p.

4.4.1.2. *Step 2*

First, it is necessary to know whether the material function has invariant properties. To do this, it is necessary to verify:

– whether the standard dimensionless material function $w(u)$ takes the following form, or whether it can be approximated by a function of the following form:

$$\phi(u) = \left(1 + \beta.u\right)^{1/\beta} \quad \text{or} \quad \phi(u) = \exp(u) \qquad [4.93]$$

– or whether the dimensional material function $s(p)$ takes the following form, or whether it can be approximated by a function of the following form:

$$s(p) = \left(A + B.p\right)^{C} \quad \text{or} \quad s(p) = \exp(A + B.p) \qquad [4.94]$$

We will see that the addition of the reference abscissa to the initial relevance list depends on the invariance properties of the dimensionless material function.

1) *Non-invariant material function*

We have seen that:

– the target internal measure is a functional of the standard dimensionless material function $w(u)$;

– the standard non-dimensionalization method based on a Taylor series of the dimensionless material function normalized by $s(p_0)$ provides the theoretical framework which enables the principles of the theory of similarity to be respected, and only in the vicinity of the reference abscissa $p = p_0$ $(u = 0)$;

– the graphical representation of the standard dimensionless material function $w(u)$ depends on the reference abscissa p_0 (e.g. see Figure 4.5).

It is precisely because of these three properties that it is necessary to *add the reference abscissa p_0 in the initial relevance list of physical quantities influencing the target variable when the material function does not have invariance properties.*

An internal measure associated with p_0 will thus be formed, once the new dimensional matrix is established and the core matrix is transformed into an identity matrix (see section 2.1.3.4). Generally speaking, the internal measure associated with p_0 can be expressed in the following form:

$$\frac{p_0}{\{\text{base}\}} \text{ with } p_0 \neq 0 \tag{4.95}$$

where {base} represents the product of the repeated physical variables (chosen to constitute the core matrix) raised to different exponents; the dimension of this product is, therefore, the same as that of p_0.

In section 4.4.3, we will see that the choice of the reference abscissa is totally free provided that $p_0 \neq 0$, but some choices are more advantageous than others insofar as they help reduce the number of internal measures (the configuration of the system).

2) *Reference-invariant material function*

When the material function is reference-invariant, the graphical representations of the standard dimensionless material function are identical whatever reference abscissa is chosen. These functions are described by very specific mathematical equations (equations [4.93] and [4.94]). From a mathematical point of view, this has a significant consequence: the standard dimensionless function could be fully described by a well-established mathematical expression, independently of reference abscissa p_0, and a set of explicit arguments could be used to characterize it.

This is why, *in the case of reference-invariant material functions, the reference abscissa p_0 must not be added to the initial relevance list of physical quantities influencing the target variable.*

Compared to non-reference-invariant material functions, the space of internal measures (i.e. the configuration of the system) is, therefore, reduced to one internal measure, the one associated with p_0 (equation [4.95]). This is an optimal situation.

4.4.1.3. *Step 3*

The third step involves adding to the initial relevance list of physical quantities influencing the target variable of the material's variable physical property $s(p)$ calculated at the reference abscissa p_0, whereby:

$$s(p_0) = s_0 \hspace{5cm} [4.96]$$

This variable $s(p_0)$ will appear in one or several of the internal measures. As a result, the user should not be surprised by the fact that, depending on the reference abscissa p_0 chosen:

– the internal measures, in which $s(p_0)$ is embedded will cover different ranges of numerical values;

– and therefore, the mathematical function linking the target internal measure to the other internal measures (the process relationship) can also change.

This remains the case even for invariant material functions, the only (and not insignificant) difference being that the space of internal measures contains one internal measure less, which is the one associated with the reference abscissa.

4.4.1.4. *Step 4*

At this stage, *the set of dimensionless numbers which appear in the argument u of the standard dimensionless material function w, except from the ratio (p/p_0)*, will be added to the initial relevance list of physical quantities influencing the target variable. This set is denoted as $\{\pi_m\}$ to differentiate it from the other internal measures.

The introduction of these dimensionless numbers comes from the standard non-dimensionalization formulation. Indeed, we saw that a Taylor series allows the approximation of the standard dimensionless material function w in the vicinity of the reference abscissa whereby:

$$w(u) = \frac{s(p)}{s(p_0)} = 1 + u + u^2.q_2 + u^3.q_3 + \sum_{k=4}^{+\infty}\left[u^k.q_k\right] \qquad [4.97]$$

where the functions q_k are defined for $k \geq 1$ whereby:

$$q_k = \frac{\left[s(p_0)\right]^{k-1}.\left(\dfrac{\partial^k s}{\partial p^k}\right)_{p=p_0}}{k!\left[\left(\dfrac{\partial s}{\partial p}\right)_{p=p_0}\right]^k} \qquad [4.98]$$

As a result, the set of dimensionless numbers $\{\pi_m\}$ to be introduced into the initial relevance list is necessarily contained in the coefficients q_k and the argument u. Since the coefficients q_k are a function of $s(p_0)$, of $\left(\dfrac{d\,s(p)}{d\,p}\right)_{p=p_0}$ and of the derivatives k^{ths} of $s(p)$, the set of dimensionless numbers $\{\pi_m\}$ is therefore contained in a single argument u.

Let us now examine the definition of the argument u:

$$u = a_0.(p-p_0) \qquad [4.99]$$

where a_0 is the coefficient of the physical property s relating to the variable p which is expressed as:

$$a_0 = \frac{1}{s(p_0)}.\left(\frac{d\,s(p)}{d\,p}\right)_{p=p_0} \qquad [4.100]$$

Equation [4.99] can be put into the following form:

$$u = a_0.p_0.\left(\frac{p}{p_0}-1\right) \quad \text{with} \quad p_0 \neq 0 \qquad [4.101]$$

Two dimensionless numbers appear in equation [4.101]:

$$\left\{ a_0.p_0; \frac{p}{p_0} \right\}$$

[4.102]

As the standard dimensionless material is approximated by a Taylor series in the vicinity of the reference abscissa (when p is close to p_0), the ratio $\dfrac{p}{p_0}$ is equal to one, and therefore $\dfrac{p}{p_0}$ must not be included in the set $\left\{ \pi_m \right\}$.

As a result, the only dimensionless number to be introduced into the initial list of relevant physical quantities influencing the target variable is:

$$\left\{ \pi_m \right\} = \left\{ a_0.p_0 \right\} \text{ where } a_0.p_0 = \frac{p_0}{s(p_0)} \cdot \left(\frac{d\,s(p)}{d\,p} \right)_{p=p_0}$$

[4.103]

When the analytical expression of the dimensional material function s(p) is known (e.g. a rheological law), it is possible to give an analytical expression of the coefficient a_0, and therefore of the dimensionless number $a_0.p_0$, directly using the expression of $\left(\dfrac{d\,s(p)}{d\,p} \right)_{p=p_0}$ (equation [4.100]). In this case, we can observe that $a_0.p_0$ is expressed by a mathematical equation involving one or more dimensionless numbers. Then, the set $\left\{ \pi_m \right\} = \left\{ a_0.p_0 \right\}$ can be replaced by this (these) dimensionless number(s).

At this stage, it is important for the user to understand that, in order to define the configuration of the material, one has the choice of using:

– *the value of dimensionless number $a_0.p_0$:*

   In this case, the dimensionless numbers responsible for the evolution of the configuration of the material are not completely visible or explicit. This first possibility nevertheless has the advantages (1) of always being applicable, even without knowing the analytical form of the dimensional material function and (2) of taking account of the fact that the configuration of the system is enlarged due to the variability of the material's physical properties. This configuration may also be used to evaluate the identity of "operating points".

– or, *the set of dimensionless numbers which forms the dimensionless number $a_0.p_0$ by algebraic combinations*:

> In this case, the set of dimensionless numbers responsible for the material configuration can be explicitly identified.

We will illustrate this using examples in section 4.4.2 and in Chapters 5 (section 5.5) and 6.

4.4.1.5. *Step 5*

After these four steps, the complete list of relevant physical quantities influencing the target variable V_{target} can be established. Its form is:

– if the material function is not reference-invariant:

$$\left\{ V_{target}, V_2, V_3, \ldots, V_m, p_0, s(p_0), \{\pi_m\} \right\} \qquad [4.104]$$

where V_{target} is the target variable, $(V_2, V_3, \ldots, V_m)$ the physical quantities involved in the dimensional analysis of an identical process in which the material's physical property s is constant;

– if the material function is reference-invariant:

$$\left\{ V_{target}, V_2, V_3, \ldots, V_m, s(p_0), \{\pi_m\} \right\} \qquad [4.105]$$

Using this list, the dimensional matrix can be written. Once the repeated variables (base) have been chosen, the complete set of internal measures $\{\pi_i\}$ responsible for variation in the target internal measure, including $\{\pi_m\}$, is formed in the same way as for a material with a constant property (see Chapter 2).

It should be noted that the set $\{\pi_m\}$ is only made up of dimensionless numbers. Since these dimensionless numbers are introduced into the list of relevant physical quantities, they will remain unchanged in the configuration of the system (except, of course, if there are rearrangements between internal measures).

4.4.2. *Examples*

We will apply the rules explained above to several examples in order to unambiguously establish the complete space of dimensionless numbers influencing the target internal measure for a process which involves a material with a variable physical property. We will distinguish between two cases, depending on whether or not there is a known analytical expression of the dimensional material function. So why make such a distinction?

Throughout this chapter, the user will have noticed that we have only dealt with physical properties $s(p)$ for which there is a known analytical expression describing the variation of s with variable p (except for example 4.3). These are non-Newtonian fluids for which a rheological law gives the variation in apparent viscosity according to the shear rate $\mu_a(\dot{\gamma})$.

However, in practice, this analytical expression is often unknown and we only have one set of i discrete data $\left(p_i, s(p_i)\right)$ to describe the variation in the material's physical property s dependent on the variable p. Using this set of discrete data, how can we:

– determine the argument u associated with the standard dimensionless material function w?

– know whether the material function w is invariant?

– establish the set of internal measures $\left\{\pi_m\right\}$?

This section (and its associated appendices) will provide the user with a number of practical tools to help answer these questions. It will also illustrate the generic nature of the standard non-dimensionalization method as a procedure which can be repeated and applied whatever the physical property being studied.

4.4.2.1. *A case where the analytical expression of the dimensional material function is known*

Here, we will examine cases of pseudoplastic and Herschel–Bulkley fluids. The user can refer to Appendix 4 for examples with Bingham and Williamson–Cross fluids.

EXAMPLE 4.1.– Pseudoplastic fluid (IV)

The dimensional material function associated with a pseudoplastic fluid (equation [4.4]) takes the following form:

$$s(p) = \mu_a(\dot{\gamma}) = k.\dot{\gamma}^{(n-1)} = (A + B.p)^C$$

where $A = 0$, $B = k$, $p = \dot{\gamma}$ and $C = n - 1$

[4.106]

It, therefore, has the form of a reference-invariant material function as described in equation [4.91][16]. As a result, the reference shear rate $\dot{\gamma}_0$ should not be added to the initial list of physical quantities influencing the target variable for a process involving a pseudoplastic fluid.

We have seen (see example 4.1, section 4.3.3.1) that the argument u of the standard dimensionless material function w associated with a pseudoplastic fluid can be expressed as follows:

$$\begin{cases} a_0 = (n-1).\dot{\gamma}_0^{-1} \\ u = a_0.(\dot{\gamma} - \dot{\gamma}_0) = \dfrac{\dot{\gamma} - \dot{\gamma}_0}{\dot{\gamma}_0}.(n-1) \end{cases}$$

[4.107]

According to equation [4.103], the dimensionless number to be introduced to the initial relevance list is $a_0.\dot{\gamma}_0$. In this example, $a_0.p_0$ can be analytically expressed:

$$a_0.\dot{\gamma}_0 = \left[(n-1).\dot{\gamma}_0^{-1}\right].\dot{\gamma}_0 = (n-1)$$

[4.108]

Equation [4.108] shows that, in the expression of the dimensionless number $a_0.\dot{\gamma}_0$, the dimensionless number n appears, which has a precise physical meaning since it is the flow index of the pseudoplastic fluid. In the

---

16 We could also have considered the associated standard dimensionless material function (equation [4.79]) which takes the form of equation [4.77] with $\beta = 1/(n-1)$ (see example 4.1 in section 4.3.4.1).

configuration of the material, it is therefore interesting to replace $a_0 \cdot \dot{\gamma}_0$ by n. This then gives:

$$\{\pi_m\} = \{n\} \tag{4.109}$$

In the end, according to the rules given, the initial list of relevant physical quantities influencing the target variable will be supplemented, in the case of a process involving a pseudoplastic fluid, by:

$$\{\mu_0, n\} \tag{4.110}$$

where μ_0 is the apparent viscosity calculated to the reference shear rate $\dot{\gamma}_0$, whereby:

$$\mu_0 = \mu_a(\dot{\gamma}_0) = k \cdot \dot{\gamma}_0^{n-1} \tag{4.111}$$

COMMENT.–

All rheological parameters which appear in the dimensional material function of a pseudoplastic fluid are not included in the set of internal measures $\{\pi_m\}$; this is the case for consistency k.

EXAMPLE 4.4.– Herschel–Bulkley fluid (I)

For a Herschel–Bulkley fluid, the dimensional material function describing the variation in the apparent viscosity with the shear rate involves three parameters and is expressed as follows:

$$s(p) = \mu_a(\dot{\gamma}) = \frac{\tau_y}{\dot{\gamma}} + k \cdot \dot{\gamma}^{n-1} \tag{4.112}$$

where τ_y is the yield stress, k is the consistency and n is the flow index.

It does not have the form of an invariant function as described by equation [4.91] or [4.92]. The reference shear rate $\dot{\gamma}_0$ must, therefore, be added to the initial list of relevant physical quantities influencing the target variable for a process involving a Herschel–Bulkley fluid.

Also in this example, $a_0 . p_0$ can be expressed analytically. By definition, this gives:

$$\begin{cases} a_0.\dot{\gamma}_0 = \dfrac{1}{\mu_0} . \left(\dfrac{d\mu_a}{d\dot{\gamma}} \right)_{\dot{\gamma}=\dot{\gamma}_0} . \dot{\gamma}_0 \\ u = a_0.\dot{\gamma}_0 . \left(\dfrac{\dot{\gamma}}{\dot{\gamma}_0} - 1 \right) \end{cases} \qquad [4.113]$$

Given that:

$$\begin{cases} \mu_0 = \mu_a(\dot{\gamma}_0) = \dfrac{\tau_y}{\dot{\gamma}_0} + k.\dot{\gamma}_0^{\,n-1} \\ \left(\dfrac{d\mu_a}{d\dot{\gamma}} \right)_{\dot{\gamma}=\dot{\gamma}_0} = -\dfrac{\tau_y}{\dot{\gamma}_0^{\,2}} + (n-1).k.\dot{\gamma}_0^{\,n-2} = \dfrac{-\tau_y + (n-1).k.\dot{\gamma}_0^{\,n}}{\dot{\gamma}_0^{\,2}} \end{cases} \qquad [4.114]$$

We can deduce that:

$$a_0 = \dfrac{1}{\mu_0} \left(\dfrac{-\tau_y + (n-1).k.\dot{\gamma}_0^{\,n}}{\dot{\gamma}_0^{\,2}} \right) \qquad [4.115]$$

We can define the dimensionless number Bi so that[7]:

$$Bi = \dfrac{\tau_y}{\mu_0 . \dot{\gamma}_0} \qquad [4.116]$$

By combining equations [4.115] and [4.116], we obtain:

$$a_0 = -\dfrac{Bi}{\dot{\gamma}_0} + \dfrac{(n-1).k.\dot{\gamma}_0^{\,n-1}}{\mu_0.\dot{\gamma}_0} = -\dfrac{Bi}{\dot{\gamma}_0} + \dfrac{(n-1).\left[\mu_0 - \dfrac{\tau_y}{\dot{\gamma}_0} \right]}{\mu_0.\dot{\gamma}_0}$$

$$= -\dfrac{Bi}{\dot{\gamma}_0} + \dfrac{(n-1).\left[1 - \dfrac{\tau_y}{\mu_0.\dot{\gamma}_0} \right]}{\dot{\gamma}_0} = -\dfrac{Bi}{\dot{\gamma}_0} + \dfrac{(n-1).\left[1 - Bi \right]}{\dot{\gamma}_0} \qquad [4.117]$$

Hence:

$$
\begin{cases}
a_0.\dot{\gamma}_0 = -Bi + (n-1).[1-Bi] \\
u = \left(-Bi + (n-1).[1-Bi]\right).\left(\dfrac{\dot{\gamma}}{\dot{\gamma}_0}-1\right)
\end{cases}
\qquad [4.118]
$$

Therefore, the analytical expression of the dimensionless number $a_0.\dot{\gamma}_0$ includes two dimensionless numbers, Bi and n, both with a known physical meaning (Bingham number and flow index). Consequently, it is relevant, in the configuration of the material, to replace $a_0.\dot{\gamma}_0$ by these two numbers. This, therefore, gives:

$$
\left\{\pi_m\right\} = \left\{a_0.\dot{\gamma}_0\right\} = \left\{n, Bi\right\}
\qquad [4.119]
$$

Finally, the initial list of relevant physical quantities influencing the target variable will be supplemented, in the case of a process involving a Herschel–Bulkley fluid, by:

$$
\left\{\dot{\gamma}_0, \mu_0, n, Bi\right\}
\qquad [4.120]
$$

COMMENTS.–

– Here also, not all the rheological parameters describing the dimensional material function of a Herschel–Bulkley fluid are included in the set of internal measures $\{\pi_m\}$. This is the case for consistency k, the yield stress. τ_y is, nevertheless, included in Bi.

– It is possible to determine the analytical expression of the standard dimensionless material function w associated with a Herschel–Bulkley fluid depending on the argument u.

Using equation [4.118], it is possible to deduce:

$$
\begin{aligned}
\dot{\gamma} &= \dot{\gamma}_0 \cdot \left[\frac{u}{-Bi + (n-1).(1-Bi)} + 1\right] \\
&= \dot{\gamma}_0 \cdot \left[\frac{u}{n.(1-Bi)-1} + 1\right]
\end{aligned}
\qquad [4.121]
$$

The standard dimensionless material function w associated with a Herschel–Bulkley fluid can be expressed as follows:

$$w(u) = \frac{\mu_a}{\mu_0} = \frac{\dfrac{\tau_y}{\dot{\gamma}} + k.\dot{\gamma}^{n-1}}{\mu_0} \qquad [4.122]$$

By inserting equation [4.121] into equation [4.122], we find:

$$w(u) = \frac{1}{\mu_0}\left[\frac{\tau_y}{\dot{\gamma}_0.\left[\dfrac{u}{n.(1-Bi)-1}+1\right]} + k.\dot{\gamma}_0^{n-1}.\left[\frac{u}{n.(1-Bi)-1}+1\right]^{n-1}\right]$$

$$= \frac{\tau_y}{\mu_0.\dot{\gamma}_0.\left[\dfrac{u}{n.(1-Bi)-1}+1\right]} + \frac{k.\dot{\gamma}_0^{n-1}}{\mu_0}.\left[\frac{u}{n.(1-Bi)-1}+1\right]^{n-1}$$

$$\qquad [4.123]$$

$$= \frac{Bi}{\dfrac{u}{n.(1-Bi)-1}+1} + \frac{\mu_0 - \dfrac{\tau_y}{\dot{\gamma}_0}}{\mu_0}.\left[\frac{u}{n.(1-Bi)-1}+1\right]^{n-1}$$

$$w(u) = \frac{Bi}{\dfrac{u}{n.(1-Bi)-1}+1} + (1-Bi).\left[\frac{u}{n.(1-Bi)-1}+1\right]^{n-1}$$

Equation [4.123] confirms that the standard dimensionless material function $w(u)$ associated with a Herschel–Bulkley fluid does not have the form of an invariant function (equation [4.77] or [4.78]). Moreover, it can be observed that the dimensionless numbers which appear in $w(u)$ are indeed the same as those identified previously in $a_0.\dot{\gamma}_0$ (equation [4.118]), namely Bi and n.

4.4.2.2. A case where there is no known analytical expression of the dimensional material function

Here, we present the example of the dependence of the viscosity of Newtonian fluids on temperature. This is a case which is commonly encountered in many thermal processes, since fluids often undergo a variation in temperature (heating and cooling). In such cases, there is no known analytical expression which describes this variation $\mu(\theta)$.

The user can refer to Appendix 5 for another example: the dependence of the surface tension of an aqueous mixture on the volume fraction of butanol.

EXAMPLE 4.5.– Dependence of the viscosity of a Newtonian fluid on temperature

In order to provide an illustration, let us consider an aqueous solution of sucrose with a concentration of 66% (w/w). Table 4.1 shows the values of viscosities μ measured at different temperatures θ.

θ (K)	μ (mPa.s)
293.15	183.17
298.15	128.15
303.15	92.35
308.15	68.30
313.15	51.69
318.15	39.93
323.15	31.42
328.15	25.13
333.15	20.40
338.15	16.78

Table 4.1. *Thermo-dependence of dynamic viscosity of an aqueous solution with a sucrose concentration of 66% (w/w)*

These discrete points (θ_i, μ_i) characterize the dimensional material function $\mu(\theta)$: a graphical representation of this is given in Figure 4.7(a).

Unlike the examples in the previous section, we do not have in this case a known analytical expression to describe $\mu(\theta)$, and we are, therefore, unable to analytically determine both the argument u of the associated standard dimensionless material function and the dimensionless number $a_0.\theta_0$.

To overcome this problem, it is necessary to *find with which mathematical function this set of discrete points (θ_i, μ_i) can be approximated*. The simplest method often involves using a polynomial of the order of n, but any other choice is possible.

In this case, an order 3 polynomial whose equation is given below[17] is adapted to accurately adjust the experimental data (R^2 = 0.9989) as shown in Figure 4.7(a).

$$\mu(\theta) = -2.86 \times 10^{-6} . \theta^3 + 2.82 \times 10^{-3} . \theta^2 - 9.26 \times 10^{-1} . \theta + 1.02 \times 10^2 \qquad [4.124]$$

From here, the derivative of the approximated function can be calculated as:

$$\frac{d\mu}{d\theta} = -3 \times 2.86 \times 10^{-6} . \theta^2 + 2 \times 2.82 \times 10^{-3} . \theta - 9.26 \times 10^{-1} \qquad [4.125]$$

Let us now fix the reference temperature θ_0, e.g. θ_0 = 303.15 K.

According to equations [4.124] and [4.125], it leads to:

$$\begin{cases} \mu_0 = \mu(303.15) = 9.26 \times 10^{-2} \quad \text{Pa.s} \\ \left(\dfrac{d\mu}{d\theta} \right)_{\theta_0 = 303.15} = -6.31 \times 10^{-2} \quad \text{Pa.s.K}^{-1} \end{cases} \qquad [4.126]$$

The argument u of the standard dimensionless material function is defined as:

$$\begin{cases} a_0 . \theta_0 = \dfrac{1}{\mu_0} . \left(\dfrac{d\mu}{d\theta} \right)_{\theta = \theta_0} . \theta_0 = -20.62 \\ u = a_0 . \theta_0 . \left(\dfrac{\theta}{\theta_0} - 1 \right) \end{cases} \qquad [4.127]$$

The standard dimensionless material function w can thus be written as:

$$w = \frac{\mu(\theta)}{\mu_0} = \frac{-2.86 \times 10^{-6} . \theta^3 + 2.82 \times 10^{-3} . \theta^2 - 9.26 \times 10^{-1} . \theta + 1.02 \times 10^2}{9.26 \times 10^{-2}} \qquad [4.128]$$

Using equations [4.127] and [4.128], it is now possible to trace the standard dimensionless material function w according to its argument u in the

---

17 In this relationship (and those that result from it), viscosity is expressed in Pa.s and temperature is expressed in K.

form of a set of discrete points (u_i, w_i). Figure 4.7(b) shows this in the form of a graph. It can be observed that the standardization conditions are clearly respected ($u = 0$), and that the standard dimensionless material function and its derivative (slope of the curve) are equal to 1 in the vicinity of $u = 0$.

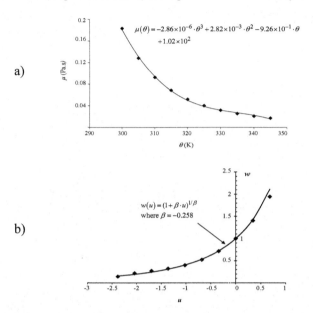

a)

b)

Figure 4.7. *Thermo-dependence of dynamic viscosity for an aqueous solution with a sucrose concentration of 66% (w/w): a) dimensional material function (continuous black line: equation [4.124]). b) Associated standard dimensionless material function (reference temperature θ_0 = 303.15K; continuous line: equation [4.130] with $\beta = -0.258$)*

How do we know whether the set of discrete points (u_i, w_i) correspond to an invariant standard dimensionless material function $\phi(u)$? If this is the case, according to what we have seen in section 4.3.4, they must be described by a function in the following form:

$$\phi(u) = \exp(u) \tag{4.129}$$

or

$$\phi(u) = (1 + \beta.u)^{1/\beta} \tag{4.130}$$

with β a constant which is not zero.

In order to verify this, it is necessary to examine whether the set of discrete points (u_i, w_i) obtained can be approximated by one of these functions. To do so, we can attempt to minimize, for instance, the relative mean error ε between the experimental discrete points and those predicted by equation [4.129] or [4.130] whereby:

$$\varepsilon = \frac{1}{N} \sum_{i=1}^{N} \left| \frac{\phi_{i,\exp} - \phi_{i,\mod}}{\phi_{i,\exp}} \right| \qquad [4.131]$$

For equation [4.130], this is performed by optimizing parameter β.

In this example, the function associated with equation [4.129] does not enable us to accurately describe the set of discrete points (u_i, w_i) (the relative mean error ε defined by equation [4.131] is greater than 70%). However, equation [4.130] leads to a value of the constant β equal to –0.258 ($\varepsilon < 4\%$). The associated graphical representation is shown in Figure 4.7(b).

Therefore, the standard dimensionless material function associated with the thermo-dependence of the viscosity of a 66% sucrose solution (w/w) can be approximated by an *invariant* function.

What are the consequences of these results for implementing the dimensional analysis in the case of a process involving a 66% sucrose solution (w/w) whose viscosity depends on temperature $\mu(\theta)$?

According to the rules stated, the initial list of relevant physical quantities influencing the target variable should be supplemented by:

$$\{\mu_0, \{\pi_m\}\} \qquad [4.132]$$

Indeed, we have demonstrated that the standard dimensionless material function associated with $\mu(\theta)$ was reference-invariant. Consequently, the reference temperature θ_0 must not be added to the initial relevance list.

According to equation [4.103], the dimensionless number $\{\pi_m\} = \{a_0.\theta_0\}$ appearing in the argument u of the standard dimensionless material function must be added to the initial list. For $\theta_0 = 303.15$ K, we calculated it (equation

[4.127]) due to the order 3 polynomial function (equation [4.124]) used to describe the set of discrete points (θ_i, μ_i).

Finally, the initial list of relevant physical quantities influencing the target variable will only be supplemented, in the case of a process involving a 66% sucrose solution (w/w) whose viscosity varies with temperature, by:

$$\{\mu_0,\ a_0.\theta_0\} \tag{4.133}$$

COMMENTS.–

– It is interesting to note that the coefficient a_0 defined by

$$a_0 = \frac{1}{\mu_0}\cdot\left(\frac{d\mu}{d\theta}\right)_{\theta=\theta_0} \tag{4.134}$$

is what is traditionally called the *temperature coefficient*, evaluated here at the reference temperature θ_0.

– If the approximated function describing the variation of the Newtonian viscosity with temperature had the form $\mu = a.\theta^b$, similar to the rheological law describing a pseudoplastic fluid (see equation [4.4], example 4.1), a would then be similar to the consistency k and b to the flow index n. The initial relevance list of physical quantities is supplemented, for pseudoplastic fluids, by $\{\mu_0,\ n\}$ (equation [4.110]); therefore, for a Newtonian fluid whose viscosity varies with temperature such as $\mu = a.\theta^b$, the initial relevance list will be supplemented by $\{\mu_0,\ b\}$ instead of $\{\mu_0,\ a_0.\theta_0\}$.

4.4.3. *Relevant choice of the reference abscissa*

At this stage, it is important to remember that:

– the process relationship is made up of one or several internal measures in which the physical property calculated at the reference abscissa, $s(p_0)$, is included;

– as a result, depending on the reference abscissa chosen p_0, the range covered by these internal measures will inevitably change and, therefore, the mathematical form of the process relationship will also change (see section 4.4.1.3);

– this remains valid in the case of reference-invariant material functions, the only (and not insignificant) difference being that the space of internal measures is reduced from one internal measure (the one associated with the reference abscissa[18]).

Moreover, it is important for the user to understand that:

– there is no a priori hypothesis governing the choice of the reference abscissa p_0. In other words, this means that *any value of p_0 can be chosen*;

– even though the choice of the reference abscissa p_0 is free, *some choices are more relevant than others* since they can help to *reduce the number of internal measures* $\{\pi_m\}$ to introduce in order to take account of the variability of the material's physical property.

We will refer back to the example of the Herschel–Bulkley fluid in order to illustrate this observation. The user can refer to Appendix 6 for examples using Bingham and Williamson–Cross fluids. In the same appendix, we also present summary tables which list, for various non-Newtonian fluids, the expressions of the dimensional material functions and the standard dimensionless material functions, the dimensionless numbers $\{\pi_m\}$ and the relevant choice of reference abscissa.

EXAMPLE 4.4.– Herschel–Bulkley fluid (II)

We have shown (see section 4.4.2.1) that the initial relevance list of physical quantities influencing the target variable must be supplemented, for a process involving a Herschel–Bulkley fluid, by:

$$\{\dot{\gamma}_0, \mu_0, n, Bi\} \tag{4.135}$$

We will demonstrate that it is relevant to choose a reference shear rate $\dot{\gamma}_0$ so that:

$$\dot{\gamma}_0 = \left(\frac{\tau_y}{k}\right)^{1/n} \tag{4.136}$$

where τ_y is the yield stress, k is the consistency and n is the flow index.

---

18 This is illustrated in the examples in Chapter 6.

In this case, the apparent viscosity calculated at the reference shear rate, μ_0, gives:

$$\mu_0 = \frac{\tau_y}{\dot{\gamma}_0} + k.\dot{\gamma}_0^{\,n-1} = \frac{\tau_y}{\left(\dfrac{\tau_y}{k}\right)^{1/n}} + k.\left(\frac{\tau_y}{k}\right)^{\frac{n-1}{n}} = 2.k^{1/n}.\left(\tau_y\right)^{\frac{n-1}{n}} \qquad [4.137]$$

The dimensionless number Bi defined in equation [4.116] becomes:

$$Bi = \frac{\tau_y}{\mu_0 \cdot \dot{\gamma}_0} = \frac{\tau_y}{\left(\dfrac{\tau_y}{\dot{\gamma}_0} + k \cdot \dot{\gamma}_0^{\,n-1}\right) \cdot \dot{\gamma}_0} = \frac{\tau_y}{\tau_y + k \cdot \dot{\gamma}_0^{\,n}}$$

$$= \frac{\tau_y}{\tau_y + k \cdot \left[\left(\dfrac{\tau_y}{k}\right)^{1/n}\right]^{n}} = \frac{\tau_y}{\tau_y + \tau_y} = \frac{1}{2} \qquad [4.138]$$

Due to this choice of reference shear rate (equation [4.136]), the dimensionless number Bi, therefore, becomes equal to a constant $\left(\dfrac{1}{2}\right)$, and it is no longer necessary to list it.

As a result, if the chosen reference shear rate $\dot{\gamma}_0$ is that given by equation [4.136], the list of relevant physical quantities influencing the target variable will be supplemented, for a process involving a Herschel–Bulkley fluid, only by:

$$\left\{\dot{\gamma}_0 = \left(\frac{\tau_y}{k}\right)^{1/n}, \ \mu_0 = 2.k^{1/n}.\left(\tau_y\right)^{\frac{n-1}{n}}, \ n\right\} \qquad [4.139]$$

Equation [4.139] is made up of one internal measure less, Bi, compared to equation [4.135] obtained for any reference shear rate $\dot{\gamma}_0$. Nevertheless, it is important to bear in mind that, even though the internal measure Bi no longer explicitly appears in equation [4.139], it is still part of the configuration of the system (star graph), but its value is fixed and equal to $\frac{1}{2}$.

4.5. Guided example 2

This final section presents a guided example which illustrates how to apply the methodology presented in this chapter in order to carry out a dimensional analysis for processes involving a material with a variable physical property. To supplement this chapter, the user can refer to Chapter 6 for other examples.

Let us consider the example of the linear frictional pressure drop induced when a fluid flows, isothermally, in laminar regime, in a circular, straight and smooth section of pipe. For a Newtonian fluid, it can be shown that the physical quantities influencing the linear pressure drop (target variable) can be expressed as follows:

$$\Delta P_L = f_1(D, v, \rho, \mu) \qquad [4.140]$$

where D is the diameter of the pipe, v is the fluid's mean velocity, ρ is its density and μ is its Newtonian viscosity.

What happens to the flow of a non-Newtonian fluid? According to the rules given in section 4.4, it is necessary to:

– conserve all the physical quantities, except viscosity μ, present in the dimensional analysis of a similar process involving Newtonian fluids. This is the initial relevance list as described in equation [4.140];

– add the reference shear rate $\dot{\gamma}_0$, except for material functions which have invariant properties;

– calculate the apparent viscosity of a fluid at the reference shear rate, $\mu_0 = \mu_a(\dot{\gamma}_0)$;

– determine the dimensionless number $\{\pi_m\} = \{a_0.p_0\}$ which appears in the argument u of the standard dimensionless material function w.

Equation [4.140], in the presence of non-Newtonian fluids, therefore becomes:

$$\Delta P_L = f_2\left(D,v,\rho,\dot{\gamma}_0,\mu_0,\{\pi_m\}\right) \qquad [4.141]$$

Subsequently, we will examine two types of non-Newtonian fluids: a pseudoplastic fluid and a Bingham fluid.

4.5.1. Pseudoplastic fluid

We have shown (see section 4.4.2.1, equation [4.110]) that, for a process which involves a pseudoplastic fluid, the initial list of relevant physical quantities influencing the target variable must be supplemented by[19]:

$$\{\mu_0,\, n\} \qquad [4.142]$$

with $\mu_0 = \mu_a(\dot{\gamma}_0) = k.\dot{\gamma}_0^{n-1}$ [4.143]

Equation [4.141] here becomes:

$$\Delta P_L = f_{pp}\left(D,v,\rho,\mu_0,n\right) \qquad [4.144]$$

where the subscript "pp" indicates that function f is associated with a pseudoplastic fluid.

The dimensional matrix **D** associated with this problem is written in Figure 4.8. The set of chosen repeated variables is { D, ρ, v }.

	ΔP_L	n	μ_0	D	ρ	v
M	1	0	1	0	1	0
L	-2	0	-1	1	-3	1
T	-2	0	1	0	0	-1

Figure 4.8. *Linear pressure drop in a pipe (laminar flow) of a pseudoplastic fluid: dimensional matrix **D** (black outline), core matrix **C** (dark gray boxes) and residual matrix **R** (light gray boxes)*

19 The reference shear rate $\dot{\gamma}_0$ is not added to the initial relevance list because the material function associated with a pseudoplastic fluid is invariant (see equation [4.80]).

The modified dimensional matrix $\mathbf{D_m}$ (obtained after transforming the core matrix into an identity matrix) is shown in Figure 4.9.

	ΔP_L	n	μ_0	D	ρ	v
	-1	0	1	0	0	0
	1	0	1	1	1	0
	2	0	1	0	0	1

Figure 4.9. *Linear pressure drop in a pipe (laminar flow) of a pseudoplastic fluid: modified dimensional matrix $\mathbf{D_m}$ (black outline), identity matrix $\mathbf{C_1}$ (dark gray boxes) and modified residual matrix $\mathbf{R_m}$ (light gray matrix)*

Finally, the dimensional analysis transforms equation [4.144] into:

$$\pi_{target} = \frac{\Delta P_L.D}{\rho.v^2} = F_{pp1}\left(\pi_2 = n,\ \pi_3 = \frac{\mu_0}{\rho.v.D}\right) \qquad [4.145]$$

The dimensionless numbers involved in equation [4.145] correspond to the coefficient of the pressure drop due to friction (f) and the inverse of the Reynolds number calculated at the reference shear rate $\dot{\gamma}_0$ (Re_0):

$$\begin{cases} \dfrac{\Delta P_L.D}{\rho.v^2} = f \\[3mm] \pi_3^{-1} = \dfrac{\rho.v.D}{\mu_0} = Re_0 \end{cases} \qquad [4.146]$$

Therefore, the dimensional analysis indicates that, during the laminar flow of a pseudoplastic fluid in a circular, straight and smooth section of pipe, there is a process relationship F_{pp2} between the coefficient of pressure drop due to friction (f) and the internal measures n and Re_0 whereby:

$$f = F_{pp2}\left(n,\ Re_0\right) \qquad [4.147]$$

In this case, the configuration of the system is increased by one internal measure (n) compared to the configuration established for a Newtonian fluid (equation [2.75]).

This result can be retrieved analytically. Indeed, when a pseudoplastic fluid flows in a laminar regime in a circular, straight and smooth section of pipe, the shear at the walls $\dot{\gamma}_W$ is analytically expressed as [MCC 93]:

$$\dot{\gamma}_W = 8.\frac{v}{D}.\frac{3n+1}{4n} \tag{4.148}$$

In this case, it is also possible to demonstrate [MCC 93] that the Fanning friction factor, equal to $f/2$, has the following analytical expression:

$$\frac{f}{2} = \frac{16}{Re_g} \tag{4.149}$$

where Re_g is a "generalized" Reynolds number defined by:

$$Re_g = 8^{1-n}.\frac{\rho.v^{2-n}.D^n}{k}.\left(\frac{4n}{3n+1}\right)^n \tag{4.150}$$

As demonstrated in section 4.4.3, the choice of reference shear rate $\dot{\gamma}_0$ is free. Let us consider that it is equal to $\dot{\gamma}_w$. The Reynolds number Re_0 thus becomes:

$$Re_0 = \frac{\rho.v.D}{\mu_0} = \frac{\rho.v.D}{k.\left(8.\dfrac{v}{D}.\dfrac{3n+1}{4n}\right)^{n-1}} \tag{4.151}$$

$$= \frac{\rho.v^{2-n}.D^n}{k.8^{n-1}.\left(\dfrac{3n+1}{4n}\right)^{n-1}} = \frac{3n+1}{4n}Re_g$$

Consequently, the Fanning friction factor has the analytical expression:

$$\frac{f}{2} = \frac{16}{\dfrac{Re_0}{\dfrac{3n+1}{4n}}} \tag{4.152}$$

The analytical solution provided by equation [4.152] includes two dimensionless numbers, Re_0 and n. This is in agreement with the configuration determined by blind dimensional analysis (equation [4.147]).

Equation [4.149] is traditionally represented in the form of a chart (Figure 4.10) representing the Fanning friction factor ($f/2$) in relation to the "generalized" Reynolds number.

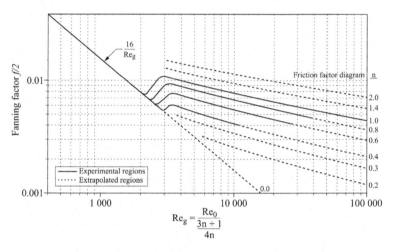

Figure 4.10. *Fanning friction factor as a function of the Reynolds number Re_g for various flow indices n [GOV 87]*

4.5.2. Bingham fluid

We have shown (Appendix 4, equation [A4.8]) that, for a process involving a Bingham fluid, the initial list of relevant physical quantities influencing the target variable must be supplemented by:

$$\{\dot{\gamma}_0, \mu_0, Bi\}$$ [4.153]

with $\mu_0 = \mu_a(\dot{\gamma}_0) = \dfrac{\tau_y}{\dot{\gamma}_0} + \mu_p$ [4.154] and Bi is the dimensionless number defined by:

$$Bi = \frac{\tau_y}{\mu_0 \cdot \dot{\gamma}_0} \qquad\qquad [4.155]$$

Equation [4.141] here becomes:

$$\Delta P_L = f_B\left(D, v, \rho, \dot{\gamma}_0, \mu_0, Bi\right) \qquad\qquad [4.156]$$

where the subscript "B" indicates that function f is associated with a Bingham fluid.

The dimensional matrix **D** associated with equation [4.156] is described in Figure 4.11. Here again, we have chosen $\{D, v, \rho\}$ as the repeated variables.

	ΔP_L	Bi	μ_0	$\dot{\gamma}_0$	D	ρ	v
M	1	0	1	0	0	1	0
L	-2	0	-1	0	1	-3	1
T	-2	0	1	-1	0	0	-1

Figure 4.11. Linear pressure drop in a pipe (laminar flow) of a Bingham fluid: dimensional matrix **D** (black outline), core matrix **C** (dark gray boxes) and residual matrix **R** (light gray boxes)

The modified dimensional matrix **D$_m$** obtained is presented in Figure 4.12.

	ΔP_L	Bi	μ_0	$\dot{\gamma}_0$	D	ρ	V
	-1	0	1	-1	0	0	0
	1	0	1	0	1	1	0
	2	0	1	1	0	0	1

Figure 4.12. Linear pressure drop in a pipe (laminar flow) of a Bingham fluid: modified dimensional matrix **D$_m$** (black outline), identity matrix **C$_I$** (dark gray boxes) and modified residual matrix **R$_m$** (light gray boxes)

The dimensional analysis, therefore, transforms equation [4.155] into:

$$\pi_{target} = \frac{\Delta P_L.D}{\rho.v^2} = F_B \left(\pi_2 = Bi, \pi_3 = \frac{\mu_0}{\rho.v.D}, \pi_4 = \frac{\dot{\gamma}_0.D}{v} \right) \qquad [4.157]$$

The dimensionless numbers involved in equation [4.157] can be rearranged to produce the coefficient of pressure drop due to friction (f) and the Reynolds number defined in the reference shear rate $\dot{\gamma}_0$ (Re_0). Equation [4.157] can, therefore, be expressed as:

$$\frac{f}{2} = F_{B1} \left(Bi, Re_0, \frac{\dot{\gamma}_0.D}{v} \right) \qquad [4.158]$$

We have seen in Appendix 6 that this space could be reduced by taking account of the reference shear rate $\dot{\gamma}_0$ so that:

$$\dot{\gamma}_0 = \frac{\tau_y}{\mu_p} \qquad [4.159]$$

Hence:

$$Bi = \frac{1}{2} \text{ and } \mu_0 = 2\mu_p \qquad [4.160]$$

Consequently, the Reynolds number Re_0 is then equal to:

$$Re_0 = \frac{\rho.v.D}{2\mu_p} \qquad [4.161]$$

The internal measure π_4, thus denoted by B_m, is defined as:

$$\frac{\dot{\gamma}_0.D}{v} = \frac{\tau_y.D}{\mu_p.v} = B_m \qquad [4.162]$$

Finally, dimensional analysis indicates that, during the laminar flow of a Bingham fluid in a circular, straight and smooth section of pipe, there is a process relationship F_{B2} between the coefficient of pressure drop due to friction (f) and the internal measures Re_0 and B_m whereby:

$$\frac{f}{2} = F_{B2}\left(Re_0 = \frac{\rho.v.D}{2\mu_p}, B_m \right)$$
[4.163]

In this case, the configuration of the system is increased by one internal measure compared to the configuration established for a Newtonian fluid (equation [2.75]).

It is possible to achieve this same result analytically. Indeed, Buckingham [BUC 21] and Hedström [HED 52] showed that, during the laminar flow of a Bingham fluid in a circular, straight and smooth section of pipe, the Fanning friction factor (equal to $f/2$) can be predicted by the following equation:

$$\frac{1}{Re_p} = \frac{f/2}{8} - \frac{1}{6}.\frac{He}{(Re_p)^2} + \frac{1}{24}.\frac{He^4}{(f/2)^3(Re_p)^8}$$
[4.164]

which can also be expressed as:

$$f = \frac{16}{Re_p}(1 + \frac{1}{6}.\frac{He}{Re_p} - \frac{1}{3}.\frac{He^4}{f^3(Re_p)^7}$$
[4.165]

where:

– Re_p is the Reynolds number

$$Re_p = \frac{\rho.v.D}{\mu_p}$$
[4.166]

– He is the Hedström number

$$He = \frac{\tau_y.\rho.D^2}{\mu_p^2}$$
[4.167]

Equation [4.165] is often graphically represented in the form of a chart [HED 52], as illustrated in Figure 4.13.

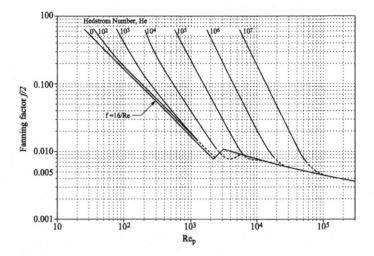

Figure 4.13. *Fanning friction factor as a function of the Reynolds number Re_p for various Hedström numbers [HED 52]*

The Reynolds number Re_p defined by equation [4.166] is the ratio between the Reynolds number Re_0 and the dimensionless number Bi when $\dot{\gamma}_0 = \dfrac{\tau_y}{\mu_p}$:

$$Re_p = \frac{Re_0}{Bi} = \frac{\dfrac{\rho.v.D}{2\mu_p}}{1/2} = \frac{\rho.v.D}{\mu_p} \qquad [4.168]$$

Concerning the Hedström number, it is the product of the dimensionless number B_m and the Reynolds number Re_p:

$$He = B_m \times Re_p = \frac{\tau_y.D}{\mu_p.v} \times \frac{\rho.v.D}{\mu_p} = \frac{\tau_y.\rho.D^2}{\mu_p{}^2} \qquad [4.169]$$

Equation [4.164] may, therefore, be expressed in the form of a process relationship linking f to B_m and Re_0. The space of internal measures produced by dimensional analysis (equation [4.162]) is, therefore, equivalent to the one given by the analytical solution of equation [4.164].

COMMENT.–

A similar approach can be taken in the case of a Herschel–Bulkley fluid. Thus, dimensional analysis shows that there is a process relationship F_{HB} between the Fanning friction factor and the four other internal measures whereby:

$$\frac{f}{2} = F_{HB1}\left(Bi, n, Re_0, \frac{\dot{\gamma}_0.D}{v} \right) \tag{4.170}$$

By choosing the reference shear rate so that:

$$\dot{\gamma}_0 = \left(\frac{\tau_y}{k} \right)^{1/n} \tag{4.171}$$

The space of dimensionless numbers described in equation [4.170] can be reduced as follows:

$$\frac{f}{2} = F_{HB2}\left(n, Re_0 = \frac{\rho.v.D}{2.k^{1/n}.\left(\tau_y\right)^{\frac{n-1}{n}}}, \frac{\left(\frac{\tau_y}{k}\right)^{1/n}.D}{v} \right) \tag{4.172}$$

Delaplace et al. [DEL 08] showed that the internal measures appearing in equation [4.172] and in the analytical solution to this problem are indeed identical.

Dimensional Analysis: A Tool for Addressing Process Scale-up Issues

This chapter aims to illustrate the advantages of dimensional analysis to address issues of process scale-up or scale-down.

To this end, we will:

– first list the principles and the rules resulting from dimensional analysis which must be respected in order to successfully complete the scale change of a process (section 5.1);

– then illustrate how to apply these rules using a number of guided examples (sections 5.2–5.5).

5.1. Conditions to satisfy to ensure complete similarity between the two scales: conservation of the operating point

5.1.1. Configuration of the system and operating points

According to the Vaschy–Buckingham theorem (see Chapter 2, section 2.1.3.3), any physical quantity representing a phenomenon (any target variable V_1 or V_{target}) function of m independent physical quantities, V_i, measured by n_d fundamental dimensions, may be described by an implicit function between $m–n_d$ dimensionless numbers π_i.

$$V_{target} = V_1 = f \left(V_2, V_3, V_4, ..., V_m \right)$$ [5.1]

thus becomes:

$$\pi_{\text{target}} = \pi_1 = F\left(\pi_2, \pi_3, \pi_4, ..., \pi_{m-n_d}\right) \qquad [5.2]$$

$\left\{\pi_i\right\}$ are the set of dimensionless ratios of physical quantities linked to the system being studied. These dimensionless numbers are obtained by dividing each non-repeated physical variable by a product of the repeated variables raised to different exponents a_{ik} (equation [5.3]) which are rational numbers and may be zero, $i \in [1; m - n_d]$ and $k \in [m - n_d + 1; m]$

$$\pi_i = \frac{V_i}{\prod_k V_k^{a_{ik}}} \qquad [5.3]$$

The set $\left\{\pi_i\right\}$ also represents the internal measures of the non-repeated physical variables in the base made up of repeated physical variables. These internal measures are independent from the units used to measure the physical quantities. Each internal measure, π_j (with $j > 1$), characterizes one of the potential causes of the variation in the target internal measure, π_{target}.

The numbers π_j can have accurate physical meanings (see Appendix 2), such as force ratios, time scale ratios, length ratios, etc.

In Chapter 2 (section 2.3.1), we referred to the *configuration of the system* as being the complete set of internal measures, $\{\pi_j$ with $\pi_j \neq \pi_{\text{target}})$, responsible for the variations in the target internal measure π_{target}. The number $(m-n_d-1)$ of internal measures which make up the configuration depends on the complexity of the system being analyzed.

According to the numerical value taken by (or imposed on) the dimensional physical quantities V_i (with $i > 1$) involved in the internal measures π_j (with $j > 1$), the dimensionless numbers π_j take different numerical values. This mathematical relationship defining the configuration of the system is therefore associated with a set of numerical values, known as *operating point*. An operating point is therefore an ordered collection of numerical values (see Chapter 2, section 2.3.1) responsible for the evolution of the target variable.

5.1.2. *Rules of similarity*

5.1.2.1. *Concept of complete similarity*

Guaranteeing the similarity of the mechanisms on two scales (laboratory-scale and industrial-scale equipments) requires the configuration of the two systems to be made up of the same internal measures and the value of the internal measures to be completely identical.

This case defines *complete* or *total similarity*.

In other words, this means that, for π_{target} to remain unchanged on two different scales, it is necessary to ensure that each of the internal measures $\pi_2, \pi_3, ..., \pi_{m-n_d}$ of both systems simultaneously take the same value on both scales.

These conditions are standardized by *similarity relationships* which are traditionally expressed as follows:

$$\text{if } \forall j > 1, \; \left(\pi_j \right)_{laboratory-scale} = \left(\pi_j \right)_{industrial-scale} = \text{idem} \qquad [5.4]$$

$$\text{then } \left(\pi_{target} \right)_{laboratory-scale} = \left(\pi_{target} \right)_{industrial-scale} = \text{idem} \qquad [5.5]$$

5.1.2.2. *Concept of partial similarity*

The concept of complete similarity as it has just been defined naturally leads to the following question: is it always possible to obtain complete similarity between the laboratory-scale equipment and the industrial-scale equipment?

We should immediately point out that it is highly likely that laboratory-scale equipment is not suitable for simulating an industrial operating point. Indeed, when the size of laboratory-scale equipment is fixed, there are occasionally not enough degrees of freedom (in terms of the operating conditions and material properties) in order to simultaneously obtain the same numerical values for the internal measures on the laboratory- and industrial-scales.

Therefore, if at least one of the numerical values of the internal measures is not exactly identical on both scales, the conditions of complete similarity are no longer fulfilled: this is a case of *partial similarity*. In this situation, it is no longer possible to guarantee that the evolution of the target internal measure, π_{target} , will be identical on both scales. It is therefore necessary to evaluate whether not conserving the numerical values of these internal measures has a significant influence on π_{target} .

It is understandable that the conditions of complete similarity become increasingly difficult to achieve as the number of physical quantities necessary to describe the system increases. Nevertheless, in any case, producing dimensionless numbers and a graphical representation of the configuration of the system provides a tool for:

– comparing the different operating points tested for the system;

– getting a better understanding of the contribution of each internal measure on the physical evolution of the system. From the knowledge of the process relationship, it will be then possible to quantify the specific influence of each internal measure on the variation of the target variable.

Any researcher using laboratory-scale equipment must therefore be able to formalize the process studied in terms of dimensionless numbers in order to identify the configuration of the system and to be able to transpose the results obtained onto another scale.

Reformulating a physical problem by establishing the relevant dimensionless numbers gives a rigorous theoretical framework to accurately ascertain the identity of the mechanisms governing the evolution of the target variable in two processes of different sizes.

5.1.2.3. *Contribution of the process relationship to the scale change*

As we will see in Chapter 6, dimensional analysis, combined with experiments on laboratory-scale equipment, offers the opportunity to establish a process relationship linking the target internal measure to the other internal measures.

To transpose a result from one scale to another, knowledge of the process relationship is not essential if there is complete similarity.

This is not the case for partial similarity. Indeed, knowledge of the process relationship can here enable defining the scale laws to respect and proposing a new operating point in order to achieve the result on a different scale. However, it is possible that not all of the dimensionless numbers values which enter into the configuration of the system remain unchanged between the two scales. In this case, it will be necessary to ensure that the process relationship used to propose the new operating point has been properly established, namely with experiments involving this new operating point. If the range of validity of the process relationship is not wide enough, experiments with different-sized laboratory equipments must be carried out to ensure that the internal measures which are not conserved do not have a significant influence on the process.

We will refer back to these points using guided examples no. 4 (section 5.3) and no. 5 (section 5.4).

5.1.2.4. Different types of similarity

Using equation [5.4], the degree of similarity between two systems is usually gauged in order to define different types of similarity. First, the definition of geometric similarity is generally given. This is commonly used, especially for sizing the laboratory equipment.

Geometric similarity means that there is a scale identity of the distances l_i at each point and in all directions between the industrial-scale equipment and the laboratory-scale equipment. In this case, a *scale factor* is introduced between both equipments, noted F_e and is defined as follows:

$$\forall i, \quad F_e = \frac{\left(l_i\right)_{industrial-scale}}{\left(l_i\right)_{laboratory-scale}} \tag{5.6}$$

where l_i represents the characteristic lengths of the industrial-scale equipment and the laboratory-scale equipment which define the geometry at both scales.

It should be noted that "laboratory-scale" or "model" are used when studying representative equipment of the industrial process investigated at a smaller scale, whereas "industrial-scale" or "prototype" refers to industrial equipment or process. Ideally, the moral is geometrically similar to the prototype, but this is not always possible.

By analogy, with the scale factor F_e which is characteristic of the geometric similarity, it is interesting to note that, when the internal measures (ratios of various physical quantities such as forces) are required to be simultaneously similar in the model and the prototype (equation [5.4]), then the set of physical quantities of the same nature involved in the definition of the system all respect the same *scale factor*, denoted by λ. This is an extremely demanding condition because the forces may be very different in nature, from among inertial forces, gravitational forces, viscous forces, interfacial forces, or pressure forces. Indeed, this condition becomes unattainable when the forces which affect the evolution of a system are too different. In such cases, the condition of complete similarity must be accepted as being unobtainable. We attempt then to conserve a partial similarity using the dimensionless numbers which have the most significant effects.

Apart from geometric similarity, the following similarities are traditionally distinguished in chemical engineering:

– *mechanical similarity*, including static similarity, kinematic similarity, and dynamic similarity:

- the principle of *static similarity* can be defined as: two geometrically similar solid bodies are statically similar if they are subject to constant forces,

- *kinematic similarity* exists when, within two geometrically similar systems, the relationship between the velocities at every point of the system is constant,

- likewise, *dynamic similarity* exists in two geometrically similar systems when the relationship between the forces at every point in the system is constant;

– *thermal similarity*: two geometrically similar systems are thermally similar when the corresponding temperature differences remain with an identical ratio, and if they are dynamically similar when they are moving;

– *chemical similarity*: two geometrically and thermally similar systems are chemically similar when the corresponding differences in concentration remain with an identical ratio, and they are dynamically similar when moving;

Two other similarities can be encountered in these specific cases, usually in the calculation of chemical reactors:

– *luminous similarity* (photochemical reactors);

– *electrical similarity* (electrochemical reactors).

Finally, section 5.5 shows that there is one last similarity: *material similarity*. This appears for processes which involve a material with a variable physical property.

5.2. Guided example 3: cooking a chicken

This first example frequently appears in the literature on dimensional analysis [ZLO 02, SZI 07]. The objective is to show that certain processes can be easily transposed because there is complete similarity between the model and prototype scales. Before examining the extrapolation itself, we will establish the configuration of the system using a traditional dimensional analysis approach.

5.2.1. *Establishing the configuration of the system*

We are examining a chicken being cooked in a preheated oven. The chicken is placed in the oven at a temperature of 20°C. The temperature of the chicken skin θ_c is considered as immediately reaching the temperature of the oven. Therefore, the resistance to heat transfer by convection and/or radiation in the oven is negligible, that is to say that the dominant (limiting) mode for heat transport in this problem is the thermal conduction within the chicken.

The target variable is the mean temperature of the flesh at the center of the chicken, denoted as θ (Figure 5.1). From a physics standpoint, this parameter alone characterizes the progress of the cooking which will give the flesh its flavor.

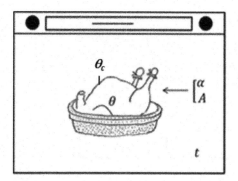

Figure 5.1. *Diagram of the physical quantities influencing the mean temperature of the flesh at the center of the chicken*

The physical quantities which have an influence on θ are:

– the *material parameters*: these are the thermal properties of the chicken flesh which can be ascertained from a single parameter: the thermal diffusivity, α. Note that α is an intermediate variable (see Chapter 3) defined as:

$$\alpha = \frac{\lambda}{\rho.c_p} \qquad\qquad [5.7]$$

where λ is the thermal conductivity of the chicken flesh, ρ is its density, and c_p is its specific heat;

– the *process parameters*: this concerns the temperature inside the oven (this is controlled by the thermostat and the oven is preheated before the chicken is placed inside). With regard to the hypotheses formulated, this temperature also dictates the temperature of the chicken skin, θ_c.

The time the chicken spends in the oven, t (i.e. the cooking time), must also be listed;

– the *boundary and initial conditions*: this concerns the exchange surface between the oven and the chicken, which is the external surface of the chicken, denoted as A.

Thus, this problem involves five physical quantities whose dimensions are expressed using the three fundamental dimensions (K, T and L). The number of associated internal measures is therefore equal to $5-3 = 2$. By taking as repeated variables $\{\theta_c, t, A\}$, the dimensional matrix is written (Figure 5.2):

	α	θ	θ_c	t	A
K	0	1	1	0	0
T	-1	0	0	1	0
L	2	0	0	0	2

Figure 5.2. *Cooking of a chicken: core matrix* **C** *(dark gray boxes), residual matrix* **R** *(light gray boxes) within the dimensional matrix* **D** *(black outline)*

By dividing the final row of this dimensional matrix by two, the core matrix becomes an identity matrix. The coefficients contained in the resulting residual matrix help to establish the set of dimensionless numbers:

$$\frac{\theta}{\theta_c} = F\left(\frac{\alpha.t}{A}\right) \tag{5.8}$$

where F is the required process relationship. The configuration of the system only contains a single internal measure ($\alpha.t/A$), which alone governs the variation in the target measure θ/θ_c:

It is interesting to note that:

– ($\alpha.t/A$) is a thermal Fourier number (see Appendix 2);

– the introduction of the intermediate variable α in the dimensional analysis approach usefully helped to reduce the configuration of the system (see Chapters 3 and 6).

5.2.2. Analysis of the configuration of the system and similarity relationship

We will now examine how the dimensional analysis carried out above can be used to achieve the same state of cooking with a larger chicken (namely to reach the same mean temperature of flesh at the center of the chicken θ in an oven at the same temperature). In particular, the objective is to show that the rules of similarity helps to determine the cooking time depending on the mass of the chicken to be cooked.

Therefore, let us imagine cooking a large chicken for eight people (prototype) in an identical way to the previous chicken for four people (model). We can assume the rather far-fetched scenario whereby the butcher is able to provide a large chicken which is both geometrically similar and whose flesh has thermal properties which are identical to those of the smaller chicken. We also assume that we are able to fit both of these chickens in our oven!

Given that the configuration of the system is described by a single internal measure (equation [5.8]), dimensional analysis provides the similarity relationships for the desired extrapolation, namely:

$$\text{if} \left(\frac{\alpha.t}{A} \right)_{model} = \left(\frac{\alpha.t}{A} \right)_{prototype} \qquad [5.9]$$

$$\text{then} \left(\frac{\theta}{\theta_c} \right)_{model} = \left(\frac{\theta}{\theta_c} \right)_{prototype} \qquad [5.10]$$

Using equation [5.9], the scale law which the cooking time must satisfy can be obtained as follows:

$$\left(\frac{\alpha.t}{A} \right) = \text{idem} \qquad [5.11]$$

such that,

$$\frac{t_{prototype}}{t_{model}} = \frac{A_{prototype}}{A_{model}} \cdot \frac{\alpha_{model}}{\alpha_{prototype}} \qquad [5.12]$$

We have assumed that the properties of the flesh are the same for the small and large chickens: $\alpha_{model} = \alpha_{prototype}$. Equation [5.7] thus becomes:

$$\frac{t_{prototype}}{t_{model}} = \frac{A_{model}}{A_{prototype}} \qquad [5.13]$$

Equation [5.13] tells us that the relationship between the cooking times must be equal to the relationship between the external surfaces of the chickens. At this stage, this information is difficult to interpret directly, because chickens are purchased according to their mass and not according to their surface area! To make the use of the previous relationship easier, it is necessary to describe the correlation between the volume and the surface for a given body:

$$m = \rho.V \equiv \rho.L^3 \qquad [5.14]$$

Given that A is identical to L^2, it becomes:

$$m \equiv \rho.A^{3/2} \qquad [5.15]$$

We can therefore conclude that:

$$A \equiv m^{2/3} \qquad [5.16]$$

Therefore, for the two chickens, large and small, which are geometrically similar and have flesh with identical thermal properties, we can thus conclude that:

$$\frac{t_{prototype}}{t_{model}} = \left(\frac{m_{prototype}}{m_{model}}\right)^{2/3} \qquad\qquad [5.17]$$

Equation [5.17] thus provides the similarity relationship which ensures an invariant target internal measure, $\pi_{target} = \theta / \theta_c$ for both scales as shown in equation [5.10].

We can immediately note that if the oven's thermostat during the cooking of both chickens is the same, then $\theta_{model} = \theta_{prototype}$ and equation [5.10] makes $\theta_{model} = \theta_{prototype} = 1$. In other words, when the relationship between the cooking times $t_{prototype}/t_{model}$ evolves proportionally to the relationship of mass to the power 2/3, the same cooking result will be obtained. This allows us, for instance, to affirm that the cooking time must increase by $(2^{2/3})$ a factor of 1.58 when the mass of the chicken is doubled.

COMMENT 5.1.– Zlokarnik [ZLO 06] notes that certain cookery books, especially American books, duly examine the cooking time for a turkey, and based on a large set of experiments, give scale laws whereby the cooking time evolves in relation to the mass to the power 0.6. This is extremely close and complies with the recommendations obtained via dimensional analysis.

COMMENT 5.2.– In this study, the prototype is obviously complex. This prevented us from fully guaranteeing the geometric similarity between the two scales (small and large chickens). Therefore, in this case, we were forced to assume that geometric similarity existed while accepting that there is a margin for error in this assumption. The fact that, in the end, the predicted conditions on the cooking of the chicken conformed with experimental results (comment 5.1) validates this assumption.

5.3. Guided example 4: power of a vertical impeller on an industrial scale

In this second example, we evaluate the behavior of an industrial installation using experiments on appropriately chosen laboratory models.

The target variable is the power consumed P by a vertical impeller on an industrial scale.

Before examining the precise rules to adopt in the experiments on the laboratory-scale equipment (or model) to reproduce the operating point on the industrial scale, we should recall the elements of dimensional analysis that we established on this system in guided example no. 1 in Chapter 2.

5.3.1. Establishing the configuration of the system

We showed that this problem could be written in terms of independent physical quantities as follows:

$$P = f\ (T, d, C_b, H_L, w, \mu, \rho, N, g) \tag{5.18}$$

where T is the diameter of the tank, d is the diameter of the impeller, C_b is the clearance between the bottom of the tank and the impeller, H_L is the height of the liquid, w the height of the impeller, N is the rotation speed of the impeller, g is the gravitational acceleration, ρ is the density of the fluid and μ is the viscosity of the Newtonian fluid (Figure 2.7).

Dimensional analysis transforms equation [5.18] as follows (see Chapter 2, section 2.4.1–2.4.4):

$$\pi_{target} = \frac{P}{\rho . N^3 . d^5}$$

$$= F\left(\begin{array}{c} \pi_2 = Re = \dfrac{\rho . N . d^2}{\mu}, \pi_3 = Fr = \dfrac{N^2 . d}{g}, \pi_4 \\[2ex] = \dfrac{T}{d}, \pi_5 = \dfrac{w}{d}, \pi_6 = \dfrac{H_L}{d}, \pi_7 = \dfrac{C_b}{d} \end{array}\right) \tag{5.19}$$

where F is the process relationship being looked for.

5.3.2. Analysis of the configuration of the system and similarity relationships

Equation [5.19] shows us that, if internal measures π_2 to π_7 take the same values (denoted as π_2 = idem, π_3 = idem, π_4 = idem, π_5 = idem and π_6 = idem) on both scales, the target variable π_1 itself retains an identical value (π_1 = idem).

The agitation system used for the experiments on the laboratory model is made up of a vertical impeller, and is defined by the following set of physical quantities:

$$\{T', d', C_b', H_L', w', \mu', \rho', N', g\} \qquad [5.20]$$

We are looking for the power P' which the impeller consumes in the laboratory model, by conducting appropriately chosen experiments. We should specify that the physical quantities which influence the power consumed on both scales (laboratory model and industrial prototype), defined in equation [5.18] and [5.20], respectively, are measured with the same units (SI).

The fluids on both scales do not necessarily need to have the same physical ($\rho \neq \rho'$) and rheological ($\mu \neq \mu'$) properties. However, we adopt the highly likely assumption that the tests are carried out on Earth and that gravitational acceleration g does not change.

The similarity relationships taking from the establishment of the dimensionless numbers (equation [5.19]) thus produce the following conditions:

if

$$\pi_2 = \frac{\rho.N.d^2}{\mu} = \frac{\rho'.N'.d'^2}{\mu'} \text{ and } \pi_3 = \frac{N^2.d}{g} = \frac{N'^2.d'}{g}$$

$$\text{and } \pi_3 = \frac{N^2.d}{g} = \frac{N'^2.d'}{g} \text{ and } \pi_4 = \frac{T}{d} = \frac{T'}{d'} \text{ and } \pi_5 = \frac{w}{d} = \frac{w'}{d'}$$

$$\text{and } \pi_6 = \frac{H_L}{d} = \frac{H_L'}{d'} \text{ and } \pi_7 = \frac{C_b}{d} = \frac{C_b'}{d'} \qquad [5.21]$$

then

$$\pi_{target} = \frac{P}{\rho.N^3.d^5} = \frac{P'}{\rho'\cdot N'^3 \cdot d'^5} \qquad [5.22]$$

Equations [5.21] and [5.22] are similarity relationships associated with this problem. Each one appears here as the similarity between two internal measures.

Equation [5.21], which expresses the constant nature of the operating point (the invariance of the numerical value of each internal measures of the system), imposes the similarity relationships which certain physical quantities must satisfy in order to ensure the invariance of the target internal measure (equation [5.22]).

Therefore, we can deduce from equation [5.21]:

1) The equality of internal measures π_4 to π_7, whereby

$$\frac{T}{T'} = \frac{d}{d'} = \frac{w}{w'} = \frac{H_L}{H'_L} = \frac{C_b}{C'_b} = F_e \qquad [5.23]$$

Equation [5.23] requires all the characteristic lengths involved in the definition of the system (T and T', d and d', w and w', H_L and H_L', C_b and C_b') to have a constant ratio. This relationship is the scale factor between the prototype and the model, F_e (equation [5.6]). The constant nature of this scale factor is essential to ensure geometric similarity between the two scales. If it had been decided that this would not be the case, for instance, that the widths do not follow the same scale factor as the heights, it would not be possible to obtain a laboratory equipment similar to the industrial agitation system. Such a distortion would prevent us from obtaining complete similarity.

2) The equality of the internal measure π_3 (Froude number), whereby

$$Fr = \frac{N^2.d}{g} = Fr' = \frac{N'^2.d'}{g} \qquad [5.24]$$

Equation [5.24] gives the similarity rule which the rotation speed of the impeller must satisfy:

$$N' = N.\left(F_e\right)^{1/2} \qquad [5.25]$$

3) The equality of the internal measure π_2 (Reynolds number), whereby:

$$Re = \frac{\rho.N.d^2}{\mu} = Re' = \frac{\rho'.N'.d'^2}{\mu'} \qquad [5.26]$$

Equation [5.26] provides the similarity rule which the properties of the fluid used in the model must satisfy in order to guarantee the invariance of the internal measure π_2 :

$$v' = \frac{\mu'}{\rho'} = \frac{\mu}{\rho} \cdot \frac{N'}{N} \cdot \frac{d'^2}{d^2} = v.F_e^{1/2}.F_e^{-2} = v.\left(F_e\right)^{-3/2} \tag{5.27}$$

COMMENT 5.3.– Therefore, by imposing on the model the operating conditions defined by equations [5.23], [5.25] and [5.27], the model and the prototype have an identical operating point. In this case, equation [5.22] is verified, and consequently, power P to use on the industrial prototype based on the power P' obtained on the model is defined as follows:

$$\frac{P}{P'} = \frac{\rho.N^3.d^5}{\rho'.N'^3.d'^5} = \frac{\rho}{\rho'}.F_e^{7/2} \cdot \tag{5.28}$$

COMMENT 5.4.– We have noted that complete similarity can be theoretically achieved for this case. However, it is necessary to ensure that the fluid being used has the following characteristics:

$$\frac{\mu}{\rho} = F_e^{3/2} \cdot \frac{\mu'}{\rho'} \tag{5.29}$$

Thus, we can note that the choice of the fluid used and the size of the model are very closely linked. In order to have more freedom in the choice of fluid to use, it is consequently worthwhile to define the scale factor according to experimentally realistic characteristics of fluids, rather than choosing the size of the model beforehand.

5.3.3. Contribution of the process relationship to the scale change

In this example, it can be observed that the behavior of the industrial equipment can be predicted with the help of experiments carried out on a model which have been carefully chosen, without knowing the exact expression of the process relationship F.

The next question which naturally arises is: what more would knowledge of the process relationship established by experiments on the laboratory scale give us?

We will demonstrate that knowledge of this relationship can provide extra opportunities for fixing operating conditions on model experiments, and therefore obtaining relevant information on the industrial process.

Through experiments on the model (see Chapter 2, guided example no. 1, equation [2.40]) we obtained the following process relationship for an agitation system with a vertical impeller centered in the tank:

$$N_P' = \frac{205.5}{Re'} \cdot \left(\frac{T'}{d'}\right)^{-0.960} \cdot \left(\frac{H_L'}{d'}\right)^{0.076} \cdot \left(\frac{C_b'}{d'}\right)^{-0.048} \tag{5.30}$$

This relationship was obtained when the internal measures defining the configuration of the system (Figure 2.11) cover the following domains $\{1.31 < T'/d' < 2.59 ; \ 0.96 < w'/d' < 1 ; \ 1.17 < H_L'/d' < 2.68 ; \ 10^{-3} < C_b'/d' < 2.44 ; \ 4.37 \times 10^{-5} < Re' < 2.47 ; 6.85 \times 10^{-9} < Fr' < 0.30\}$.

The process relationship [5.30] indicates that the power number N_P which is independent of the Froude number when it is between 685×10^{-9} and 0.30 In this case, the power number only depends on the Reynolds number and certain internal geometric measures (T'/d', H_L'/d', C_b'/d'). The influence of the internal measure w'/d' could not be evaluated because it remained almost unchanged throughout the tests (in the vicinity of 1).

In other words, if the numerical values of the internal measures involved in the prototype satisfy $\{1.31 < T/d < 2.59; \ 0.96 < w/d < 1; \ 1.17 < H_L/d < 2.68; \ 10^{-3} < C_b/d < 2.44; 4.37 \times 10^{-5} < Re < 2.47; 6.85 \times 10^{-9} < Fr < 0.30\}$, it is possible to use equation [5.30] to find conditions of invariance for N_P on both scales. Although the Froude number does not have an effect on the process relationship, it is still necessary to verify at the end of the procedure that the values of this number for the experiments on the model coincide with those produced by the prototype.

We were looking to find the power consumed by a vertical impeller with ($d = w = 0.168$ m; $C_b = 0.049$ m) placed in a tank with a diameter of $T = 0.345$ m containing a fluid with a height of $H_L = 0.3765$ m. The fluid to be agitated (between 30 and 100 tr.min^{-1}) was a glucose syrup with a density and viscosity of 1419.7 kg.m^{-3} and 163.1 Pa.s (at 19.2°C), respectively. Consequently, the numerical values of the internal measures used in the prototype, $\{T/d = 2.05; \quad w/d = 1; \quad H_L/d = 2.24; \quad C_b/d = 0.291; 0.12 < Re < 0.4; 0.025 < Fr < 0.28\}$, are in agreement with the conditions to retain the process relationship [5.30], $\{1.31 < T/d < 2.59; \ 0.96 < w/d < 1; \ 1.17$

$<H_l/d< 2.68$; $10^{-3}<C_t/d< 2.44$; $4.37\times10^{-5}<Re<2.47$; $6.85\times10^{-9}<Fr<0.30\}$: it is therefore possible to predict the values of N_p in the prototype from a process relationship established using a model which was four times smaller ($F_e = \dfrac{0.168}{0.0405} = 4.14$). When applying the relationship [5.30] to the prototype, we obtain $N_p = \dfrac{116.2}{Re}$, and consequently, $P = 116.2\times\mu.N^2.d^3$. The power consumption measured in the prototype gave us $N_p = \dfrac{106.4}{Re}$, hence $P = P = 106.4\times\mu.N^2.d^3$ (Figure 5.3).

The slight difference between the prediction and the model can be explained by the fact that the model and prototype were not perfectly geometrically similar. Indeed, the prototype tank had a curved-bottom (compared to a flat bottom on the model) to make it easier to empty. The way in which the impeller was attached to the agitation rod was also slightly different.

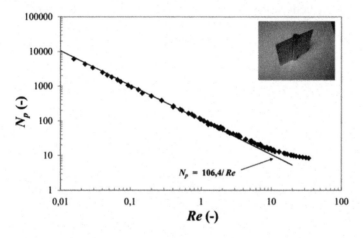

Figure 5.3. *Power curve from the test obtained with a vertical impeller on a prototype scale*

Let us now use the process relationship. With geometric similarity between the model and the prototype, the following similarities can be guaranteed:

$$\frac{T}{d} = \frac{T'}{d'} = \text{idem and } \frac{w}{d} = \frac{w'}{d'} = \text{idem and}$$

$$\frac{H_L}{d} = \frac{H'_L}{d'} = \text{idem and } \frac{C_b}{d} = \frac{C'_b}{d'} = \text{idem} \qquad [5.31]$$

The similarity relationships issued from the process relationship can therefore be summarized as follows:

$$\text{if } Re = \frac{\rho.N.d^2}{\mu} = \frac{\rho'.N'.d'^2}{\mu'} = Re' \qquad [5.32]$$

$$\text{then } N_p = \frac{P}{\rho.N^3.d^5} = \frac{P'}{\rho'\cdot N'^3 \cdot d'^5} = N'_p \qquad [5.33]$$

The equality [5.32] provides the similarity rule (equality of Reynolds numbers on both scales) to be respected by the model to guarantee the invariance of the target internal measure, N_p:

$$\frac{N}{N'} = F_e^{-2} \cdot \frac{\left(\dfrac{\mu}{\rho}\right)}{\left(\dfrac{\mu'}{\rho'}\right)} \qquad [5.34]$$

There are several solutions for ensuring the equality of Reynolds numbers on both scales (equation [5.32]):

– only adjust the properties of the fluid by changing its nature between the experiments on the model and the prototype;

– only adjust the rotation speed of the impeller;

– adjust the rotation speed of the impeller and the properties of the fluid simultaneously.

Let us assume that we are using the same fluid in the prototype and the model, and that, for reasons of flexibility, we are only adjusting the rotation speed of the impeller. Equation [5.34] can therefore be simplified as follows:

$$N = F_e^{-2} \cdot N' \qquad [5.35]$$

In this case:

$$Fr = \frac{N^2.d}{g} = \frac{\left(F_e^{-4}.F_e\right)\cdot\left(N'^2.d'\right)}{g} = F_e^{-3}.Fr' \qquad [5.36]$$

If Fr is between 6.85×10^{-9} and 0.30, equation [5.30] can be applied. Power P for the prototype is thus predicted according to the measure of power P' obtained from the model, as per equation [5.33]:

$$\frac{P}{P'} = \frac{F_e^5}{F_e^6} = F_e^{-1} \qquad [5.37]$$

Finally, it can be observed that by imposing the following operating conditions on the model:

$$T' = \frac{T}{F_e}, \ d' = \frac{d}{F_e}, \ C'_B = \frac{C_B}{F_e}, \ H'_L = \frac{H_L}{F_e}, \ w' = \frac{w}{F_e}$$

and $N' = N.F_e^2$, with $\rho = \rho'$ and $\mu = \mu'$ \qquad [5.38]

The model and the prototype have a different operating point due to the value of $Fr' = F_e^3.Fr$.

Knowledge of the process relationship established at the scale of the model nevertheless helps to highlight that, if Fr remains between 6.85×10^{-9} and 0.30, it has no influence on N_p, and thus that the power P to be used on the prototype can be calculated using equation [5.37], where $P = F_e^{-1}.P'$. However, it is impossible to access this consumed power if Fr is outside of this range, since the identity of operating points on the model and the prototype are no longer respected (equation [5.36]).

This example shows once again that guaranteeing geometric similarity between the model and the prototype and fixing identical operating conditions for physical quantities on both scales are not sufficient conditions for correctly reproducing the mechanisms observed on the prototype. It can also be observed that it is necessary to identify the industrial operating point to correctly define the experiments on the model and eventually use them to better understand the operation of the prototype.

5.4. Guided example 5: emulsification process in an agitation tank

The objective of this third example is to evaluate which operating conditions should be imposed on an industrial installation (prototype) in order to reproduce an emulsification process obtained by experiments on a laboratory scale. We should point out straight away that this process is a case of *partial similarity*. Indeed, we will see that, given the significant number of physical quantities involved (two-phase mixture, geometry of the agitation system) and the low number of parameters which can be arbitrarily chosen, it is not possible to obtain enough degrees of freedom to ensure the identity of the operating point between the model and the prototype.

Before examining the procedure to adopt in order to identify the dimensionless numbers which have a significant effect on the target internal measure, we will use dimensional analysis to establish the configuration of the system.

5.4.1. *Determining the configuration of the system*

In this case, we are examining an emulsification process in a flat-bottomed agitation tank with no baffles, and equipped with a high shear disk impeller vertically centered.

The emulsion only contains two phases (oil and water). Initially (before the impeller rotates), the two phases are spatially layered, with the aqueous (denser) phase at the bottom of the tank (Figure 5.4). The rotation of the impeller leads to the formation of an emulsion: droplets of oil are dispersed within the aqueous phase. The size of the droplets is measured as a function of time. Subsequently, it is assumed that the droplets follow a singular distribution mode and that the Sauter mean diameter measured (d_{32}) is an important indicator for this size distribution. Thus, it will be chosen as the target variable.

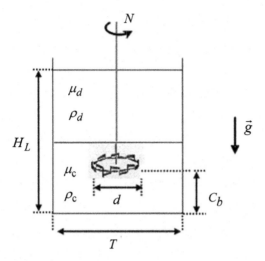

Figure 5.4. *Diagram and identification of physical quantities influencing an emulsification process in an agitation tank with a vertically centered high shear disk impeller*

The physical quantities which influence the Sauter mean diameter of droplets are:

– *material parameters*: these are the continuous phase (index c) and dispersed phase (index d) properties, which are their densities (ρ_c, ρ_d) and viscosities (μ_c, μ_d).

It is also a necessary to list the volume fraction, φ, which is the ratio between the volumes occupied by the dispersed phase and the continuous phase, respectively.

Finally, the interfacial tension σ must be introduced since the phases are immiscible;

– *process parameters*: these are the rotation speed of the impeller N and the duration of the agitation, *t;*

– *boundary and initial conditions*: the complete description of the boundaries of the flow domain require a number of geometric parameters, other than T (diameter of the tank), d (diameter of the impeller), C_b (clearance between the bottom of the tank and the impeller), H_L (initial height of the two phases in the tank) (Figure 5.5). These are above all the parameters which define the geometry of the impeller (shape, size and number of teeth on the periphery of the impeller turbine). Subsequently, $\{p_{geo}\}$ will be used to account for the geometric parameters, other than T, d, C_b and H_L, necessary for the complete description of the agitation system.

It is necessary to list gravitational acceleration g because the gravitational field acts on the free surface.

In terms of independent physical quantities, the problem is therefore written as:

$$d_{32} = f(g, T, H_L, C_b, d, \{p_{geo}\}, \varphi, \sigma, \mu_c, \mu_d, \rho_c, \rho_d, N, t) \qquad [5.39]$$

The dimensional matrix which corresponds to the problem is represented in Figure 5.5. The base chosen must be made up of three repeated variables (since three fundamental dimensions are necessary to express the dimensions of the quantities listed in equation [5.39]). As per guided example no. 1 in Chapter 2, we have chosen $\{\rho_c, N, d\}$. Indeed, this base, which complies with the principles given in Chapter 2, is traditionally used in agitation/mixing problems: it was therefore logically more preferable than others.

	d_{32}	g	T	H_L	C_b	μ_c	μ_d	φ	σ	ρ_d	t	$\{p_{geo}\}$	ρ_c	d	N
M	0	0	0	0	0	1	1	0	1	1	0	0	1	0	0
L	0	1	1	1	1	-1	-1	0	0	-3	0	1	-3	1	0
T	1	-2	0	0	0	-1	-1	0	-2	0	1	0	0	0	-1

Figure 5.5. *Droplet size during an emulsification process with a high shear disk impeller: core matrix* **C** *(dark gray boxes), residual matrix* **R** *(light gray boxes) within the dimensional matrix* **D** *(black outline)*

Figure 5.6 shows the modified dimensional matrix obtained following the transformation of the core matrix into an identity matrix.

	d_{32}	g	T	H_L	C_b	μ_c	μ_d	φ	σ	ρ_d	t	$\{p_{geo}\}$	ρ_c	d	N
M	0	0	0	0	0	1	1	0	1	1	0	0	1	0	0
3M+L	1	1	1	1	1	2	2	0	3	0	0	1	0	1	0
-T	0	2	0	0	0	1	1	0	2	0	-1	0	0	0	1

Figure 5.6. *Droplet size during an emulsification process with a high shear disk impeller: identity matrix* **C₁** *(dark gray boxes), modified residual matrix* **R**$_m$ *(light gray boxes) within the modified dimensional matrix* **Dm** *(black outline)*

The coefficients contained in the modified residual matrix help to establish the space of internal measures associated with the problem in the base $\{\rho_c, N, d\}$:

$$\pi_{target} = \frac{d_{32}}{d} =$$

$$F_1 \begin{pmatrix} \pi_2 = \dfrac{g}{N^2.d}, \ \pi_3 = \dfrac{T}{d}, \pi_4 = \dfrac{H_L}{d}, \pi_5 = \dfrac{C_b}{d}, \pi_6 = \dfrac{\mu_c}{\rho_c.d^2.N}, \\[2mm] \pi_7 = \dfrac{\mu_d}{\rho_c.d^2.N^2}, \pi_8 = \varphi, \pi_9 = \dfrac{\sigma}{\rho_c.N^2.d^3}, \pi_{10} = \dfrac{\rho_d}{\rho_c}, \pi_{11} = N.t, \{\pi_{geo}\} \end{pmatrix} \quad [5.40]$$

where $\{\pi_{geo}\}$ represents the set of geometric internal measures associated with the geometric parameters $\{p_{geo}\}$.

By carrying out rearrangements between dimensionless numbers (see Chapter 2, section 2.1.3.5), the following dimensionless numbers are obtained:

– the Reynolds number for the agitation ($Re = \dfrac{\rho_c.N.d^2}{\mu_c}$) evaluated using the properties of the continuous phase, by raising π_6 to the power -1;

– the Froude number ($Fr = \dfrac{N^2.d}{g}$), by raising π_2 to the power -1;

– the viscosities ratio $\dfrac{\mu_d}{\mu_c}$, by dividing π_7 by π_6;

– the Weber number $We = \dfrac{\rho_c\, N^2 d^3}{\sigma}$ also evaluated using the properties of the continuous phase, by raising π_9 to the power -1.

These rearrangements transform equation [5.40] as follows:

$$
\pi_{target} = \frac{d_{32}}{d} = F_2\left(
\begin{array}{l}
Fr = \dfrac{N^2.d}{g},\ \dfrac{T}{d},\dfrac{H_L}{d},\dfrac{C_b}{d},\ Re = \dfrac{\rho_c.N.d^2}{\mu_c},\dfrac{\mu_d}{\mu_c},\varphi, \\[2ex]
We = \dfrac{\rho_c.N^2.d^3}{\sigma},\dfrac{\rho_d}{\rho_c}, N.t, \{\pi_{geo}\}
\end{array}
\right)
\qquad [5.41]
$$

According to equation [5.41], at least 10 internal measures define the configuration of the system, and are thus responsible for the variations in the target internal measure d_{32}/d, hence confirming the complexity of the emulsification process.

Let us look in more detail at the configuration of the system obtained.

The physical analysis of the problem shows that the size of the droplets is the result of an equilibrium between the external forces acting to break-up the droplets (forces of inertia, viscosity and gravity) and cohesive forces that maintain them (interfacial forces). It is therefore logical to find as internal measures in the configuration of the system the ratios of these forces: *We, Re, Fr*.

In the configuration of the system, we also find the internal measures which affect hydrodynamics:

1) the volume fraction of the dispersed phase (φ), the ratio of the viscosities (μ_d/μ_c) and of the densities (ρ_d/ρ_c) of both phases. These internal measures act on the inertial forces or create extra buoyancy forces (sedimentation/flotation);

2) the system's geometric internal measures: $T/d,\ H_L/d,\ C_b/d,\ \{\pi_{geo}\}$;

3) the internal measure $N.t$ which shows the influence of the duration of the agitation (i.e. the time-scale these forces act on the droplets); since the formation of these droplets is the result of a break-up and coalescence equilibrium, it is thus logical to find this ratio.

5.4.2. *Analysis of the configuration of the system and similarity relationships*

Since the configuration of the system is defined (equation [5.41]), is it possible to reproduce an operating point obtained on a model onto an industrial prototype? The answer is *"no!"*. To demonstrate this, we will only show that it is not possible to simultaneously conserve on both scales two numerical values for the internal measures in the configuration of the system, namely the values of the Reynolds and Weber numbers.

Subsequently, we will create the relevance list of physical quantities associated with the industrial prototype as follows:

$$(g, T, H_L, C_b, d, \{p_{geo}\}, \varphi, \sigma, \mu_c, \mu_d, \rho_c, \rho_d, N, t) \qquad [5.42]$$

And the relevance list of physical quantities associated with the laboratory model as:

$$(g, T', H'_L, C'_b, d', \{p'_{geo}\}, \varphi', \sigma', \mu'_c, \mu'_d, \rho'_c, \rho'_d, N', t') \qquad [5.43]$$

In this case, the emulsification is carried out with the same ingredients on both scales, whereby:

$$\varphi = \varphi', \ \sigma = \sigma', \ \mu_c = \mu'_c, \ \mu_d = \mu'_d, \ \rho_c = \rho'_c, \ \rho_d = \rho'_d \qquad [5.44]$$

And consequently:

$$\frac{\mu'_d}{\mu'_c} = \frac{\mu_d}{\mu_c} \ \text{and} \ \frac{\rho'_d}{\rho'_c} = \frac{\rho_d}{\rho_c} \qquad [5.45]$$

Furthermore, on the industrial scale, we will use a geometrically similar impeller to the one used in the model. The scale factor F_e between the prototype and the model is:

$$F_e = \frac{d}{d'} \qquad [5.46]$$

In this case, we have:

$$\left[\frac{T'}{d'} = \frac{T}{d}\right] \text{ and } \left[\frac{C_b'}{d'} = \frac{C_b'}{d'}\right] \text{ and } \left[\frac{H_L'}{d'} = \frac{H_L}{d}\right] \text{ and } \left\{\pi_{geo}'\right\} = \left\{\pi_{geo}\right\} \qquad [5.47]$$

According to the configuration established (equation [5.41]), the existence of geometric similarity (equations [5.46]–[5.47]), and the use of the same formulation produced for the prototype and the model (equation [5.44]), the following system of equations [5.48]–[5.51] must be satisfied in order to obtain an equality of operating points on both scales (conservation of the same value of the target internal measure d_{32}/d):

$$Fr = \frac{N^2.d}{g} = \frac{N'^2.d'}{g} = Fr' \text{ and} \qquad [5.48]$$

$$Re = \frac{\rho_c.N.d^2}{\mu_c} = \frac{\rho_c.N'.d'^2}{\mu_c} = Re' \text{ and} \qquad [5.49]$$

$$We = \frac{\rho_c.N^2.d^3}{\sigma} = \frac{\rho_c.N'^2.d'^3}{\sigma} = We' \text{ and} \qquad [5.50]$$

$$N.t = N'.t'' \qquad [5.51]$$

From here, it can be observed that simultaneous equality on both scales of the Reynolds and Weber numbers (rules of similarity) is impossible. Indeed, the equality of the Reynolds numbers (equation [5.49]) means that:

$$N = F_e^{-2}.N' \qquad [5.52]$$

whereas the equality of the Weber numbers (equation [5.50]) means that:

$$N = F_e^{-3/2}.N' \qquad [5.53]$$

Consequently, *partial similarity* is the only possibility.

It should be noted that:

– the equality of Froude numbers on both scales (equation [5.48]) also means that:

$$N = F_e^{-1/2}.N' \qquad [5.54]$$

This confirms the impossibility of simultaneously conserving the Reynolds, Froude and Weber numbers on both scales.

– however, conserving the internal measure connected with the duration of the agitation (equation [5.51]) on both scales is not problematic. Indeed, the agitation duration t to impose on the industrial prototype can be easily adjusted once the relationship of rotation speeds between the prototype and the model N/N' is fixed:

$$t = t'.\frac{N'}{N}$$

[5.55]

We can go further with the analysis and accurately ascertain the evolution of the operating point according to the rotation speed chosen for the prototype:

– case no. 1: $N = F_e^{-2}.N'$ (equality of Reynolds numbers)

hence:

$$Re = Re', \ Fr = F_e^{-3}.Fr' \text{ and } We = F_e^{-1}.We'$$

[5.56]

– case no. 2: $N = F_e^{-3/2}.N'$ (equality of Weber numbers)

hence:

$$We = We', \ Fr = F_e^{-2}.Fr' \text{ and } Re = F_e^{1/2}.Re'$$

[5.57]

– case no. 3: $N = F_e^{-1/2}.N'$ (equality of Froude numbers)

hence:

$$Fr = Fr', \ We = F_e^{2}.We' \text{ and } Re = F_e^{3/2}.Re'$$

[5.58]

These calculations clearly show that it would be impossible to conserve the same value for the target internal measure, d_{32}/d, on both scales:

$$\frac{\left(\dfrac{d_{32}}{d}\right)}{\left(\dfrac{d'_{32}}{d'}\right)} \neq 1$$

[5.59]

Moreover, it should be noted that in this problem, the objective is to obtain the same mean droplet size on both scales, and not the same value of the target internal measure, d_{32}/d. Indeed, if that were the case, the diameter of the droplets on an industrial scale would be directly linked to the scale factor:

$$\frac{\left(\dfrac{d_{32}}{d}\right)}{\left(\dfrac{d'_{32}}{d'}\right)} = 1 \;\rightarrow\; \frac{d_{32}}{d'_{32}} = \frac{d'}{d} = F_e \qquad\qquad [5.60]$$

5.4.3. Contribution of the process relationship in the scale change

Given this case of partial similarity, how can the droplet size be conserved on both scales? In other words, which is the most important dimensionless number to conserve?

We will answer these questions below. To do so, we will examine the process relationship obtained in the literature. There are several studies involving an agitation tank under various operating conditions with the objective of obtaining the semi-empirical correlations governing the mean diameter of the droplets. Zhou and Kresta [ZHO 98] and Xuereb et al. [XUE 06] summarized the main correlations encountered and the investigation domains explored. We note that they are generally established for Rushton-type agitators in turbulent regime. For instance, we can cite:

– the Calderbank correlation [CAL 58] for a four-blade impeller

$$\frac{d_{32}}{d} = 0.06 \times (1 + 3.75 \times \varphi) \, . \, We^{-3/5} \qquad\qquad [5.61]$$

– the Brown and Petit correlation [BRO 70] for a six-blade Rushton impeller

$$\frac{d_{32}}{d} = 0.05 \times (1 + 3.14 \times \varphi) \, . \, We^{-3/5} \qquad\qquad [5.62]$$

The examination of various correlations clearly shows that the Weber number has a major influence on the mean diameter of the droplets, as well as the volume fraction. Indeed, most correlations for fixed geometric and material parameters and composition are expressed in the following form:

$$\frac{d_{32}}{d} = C \cdot We^{-3/5} \qquad\qquad [5.63]$$

where C is a constant.

Let assume that this type of relationship remains true for high shear disk impellers, that the model and prototype are geometrically similar and involve the same material parameters. It is therefore necessary, in order to obtain a droplet diameter d_{32} in the prototype equal to the one obtained in the model d'_{32}, that:

$$\frac{(d_{32})}{(d_{32})'} = \frac{(We)^{-3/5}.d}{(We')^{-3/5}.d'} = \frac{(N^2.d^3)^{-3/5}.d}{(N'^2.d'^3)^{3/5}.d'} = 1 \qquad\qquad [5.64]$$

By introducing the scale factor $F_e = d/d'$, equation [5.53] becomes:

$$\left(\frac{N}{N'}\right)^{-6/5}.F_e^{-9/5}.F_e = 1 \qquad\qquad [5.65]$$

hence:

$$N = F_e^{-2/3}.N' \qquad\qquad [5.66]$$

where N and N' are the rotation speeds of the impeller in the prototype and the model, respectively.

Therefore, knowledge of the process relationship established at the model scale is a key element for defining the operating conditions to adopt at the industrial scale in order to attempt to obtain the same droplet size as the one obtained in the model, namely:

$$T = \frac{T'}{F_e}, \quad d = \frac{d'}{F_e}, \quad C_b = \frac{C_b'}{F_e}, \quad H_L = \frac{H_L'}{F_e}, \quad w = \frac{w'}{F_e}, \quad \varphi = \varphi', \ \sigma = \sigma',$$

$$\mu_c = \mu_c', \qquad \mu_d = \mu_d', \qquad \rho_c = \rho_c', \qquad \rho_d = \rho_d', \qquad N = F_e^{-2/3} N'$$

$$\text{and } t = F_e^{2/3}.t' \qquad\qquad [5.67]$$

However, it is important to bear in mind that, even by imposing the operating conditions defined in equation [5.67], some internal measures associated with the prototype's operating point are relatively distant from their initial value in the model, given that:

$$We = F_e^{5/3} . We' \text{ and } Fr = F_e^{-1/3} Fr' \text{ and } Re = F_e^{4/3} . Re' \qquad [5.68]$$

In order to evaluate the consequences of equation [5.68] on the diameter of droplets obtained in the prototype, it is essential to verify that the process relationship used as a basis for our reasoning (equation [5.63]) has been established within the ranges of the Weber, Froude and Reynolds numbers reached in the prototype. The displacement of the operating point in the prototype compared to the operating point in the model may thus induce changes in dominant mechanisms and therefore some modifications in the break-up/coalescence process of droplets.

5.5. Specific case of a scale change in a process involving a material with a variable physical property (guided example 6)

5.5.1. Preamble

Chapter 4 showed that the number of internal measures characterizing the configuration of the system increases in the presence of a material with a variable physical property: the *configuration of the material* changes and as a consequence, so does the complete configuration of the system including the set of physical quantities in the process.

This new configuration of material may be established unambiguously by using the notion of a standard dimensionless material function. Consequently, in the case of process involving a material with a variable physical property, the rule of similarity remains the same to ensure complete similarity between the industrial prototype and the model: it is necessary to obtain the identity of the operating point on both scales. As with any transposition, this requires the numerical values of the set of internal measures defining the configuration of the system to be simultaneously conserved, including the set of internal measures associated with the variability of the material's physical property $\{\pi_m\}$. We will illustrate this using a final guided example.

5.5.2. Guided example 6: scaling of a scraped surface heat exchanger

Here we will use the example provided by Pawlowski [PAW 91], dealing with the dimensions of a scraped surface heat exchanger. The objective is to illustrate how the approach of transposing the results must be carried out when the process involves a material with a variable physical property. Above all, it is necessary to show how the model material and/or the reference abscissa must be chosen to achieve complete similarity when a different material is used on both scales.

5.5.2.1. Description of the problem

The situation is as follows: an industrial scraped surface heat exchanger must be scaled in order to heat a product with Newtonian behavior flowing at a volumetric flow rate Q from an inlet temperature θ_{in} to an outlet temperature θ_{out}. The wall temperature, θ_w, of the scraped surface exchanger is also industrially imposed by the transformation process.

More precisely, the industrial specifications to comply with are as follows:

– input product of the temperature: $\theta_{in} = 102°C$

– output temperature of the product: $\theta_{out} = 112°C$

– temperature of the wall: $\theta_w = 118°C$

– volumetric flow rate of the product: $Q = 1.33 \times 10^{-4}\,m^3.s^{-1}$

The material function is the dependence of the product's viscosity on the temperature, $\mu(\theta)$. Pawlowski [PAW 91] verified that in the range of temperatures covered, the physical properties other than viscosity can be considered as independent of temperature.

Pawlowski [PAW 91] showed that the standard dimensionless material function of the product can be described by a reference-invariant standard function. However, the theoretical mathematical relationship linking viscosity to temperature is not known (see Chapter 4, section 4.4.2.2).

Consequently, physical quantities which influence the target variable must be added to the initial relevance list:

– the viscosity of the product calculated at a reference temperature θ_0, arbitrarily chosen: $\mu(\theta_0) = \mu_0$;

– the dimensionless number, $\pi_m = a_0.\theta_0$, which appears in the argument u of the standard dimensionless material function w except the ratio (θ / θ_0), whereby:

$$a_0.\theta_0 = \frac{\theta_0}{\mu_0} \cdot \left(\frac{d\mu(\theta)}{d\theta} \right)_{\theta=\theta_0} \qquad [5.69]$$

As mentioned in Chapter 4 (section 4.4.1), Pawlowski [PAW 91] states this rule differently and recommends listing a_0 instead of $a_0.\theta_0$. These two approaches are exactly equivalent. However, since we have taken his example, we have kept his method. In the case when viscosity depends on temperature, a_0 is traditionally known as the temperature coefficient and denoted as γ_0 (see equation [4.44]).

The reference temperature, θ_0, is not added to the relevance list because the material function is reference-invariant. Nevertheless, its value must be chosen since it is necessary for the calculation of μ_0 and other physical properties. Pawlowski [PAW 91] considers it to be equal to the mean temperature between the two extreme temperatures encountered in the industrial installation, whereby:

$$\theta_0 = \frac{\theta_{in} + \theta_w}{2} = 110°C \qquad [5.70]$$

The physical properties of the product, calculated at this reference temperature are:

– dynamic viscosity: $\mu_0 = 0.84$ Pa.s

– temperature coefficient of viscosity: $\gamma_0 = 0.083$ K^{-1}

– density of the product: $\rho_0 = 897$ kg.m^3

– specific heat of the product: $c_{p0} = 1570$ J.kg^{-1}.K^{-1}

– thermal conductivity of the product: $\lambda_0 = 0.29$ W.m^{-1}.K^{-1}

Finally, Pawlowski [PAW 91] used one piece of equipment which was already available as a reduced model in order to limit supplementary investment, therefore fixing the size of the model: $D' = 0.05$ m. Consequently, the geometry of the industrial equipment will be built to be identical to the model.

The scaling of the industrial prototype requires the diameter D of the equipment to be defined, as well as the rotation speed of the rotor, N, to be used to raise the temperature of the product. It is also necessary to ascertain the mechanical power P to be applied so that the rotor can be rotated at a speed of N.

The recommended approach consists of:

– firstly fixing the operating conditions of the model (nature of the fluids to use) to be able to achieve similarity with the industrial scale (prototype);

– then conducting experiments on the model (rotation speed and power associated), to predict the physical quantities to impose on the prototype.

5.5.2.2. Relevance list of independent physical quantities and establishment of the dimensionless numbers

In this problem, the target variables are the output temperature of the product, θ_{out}, and the mechanical power P to apply to turn the rotor. Using the elements shown previously in this chapter and in Chapter 4, it is possible to establish the relevance list of independent physical quantities influencing the two target variables:

$$\theta_{out} = f_{\theta 1}\left(D, Q, N, g, \theta_{in}, \theta_w, \mu_0, \gamma_0, \rho_0, c_{p0}, \lambda_0\right) \qquad [5.71]$$

$$P = f_{P1}\left(D, Q, N, g, \theta_{in}, \theta_w, \mu_0, \gamma_0, \rho_0, c_{p0}, \lambda_0\right) \qquad [5.72]$$

where g is gravitational acceleration, $f_{\theta 1}$ and f_{P1} represent the dimensional functions linking the target variables θ_{out} and P, respectively to the physical quantities of influence.

These dimensional spaces can be immediately reduced to 11 physical quantities by introducing the following intermediate variables: $\theta_{in} - \theta_w$ and $\theta_{out} - \theta_w$. Thus, equations [5.71] and [5.72] become:

$$\left(\theta_{out} - \theta_w\right) = f_{\theta 2}\left(D, Q, N, g, \left(\theta_{in} - \theta_w\right), \mu_0, \gamma_0, \rho_0, c_{p0}, \lambda_0\right) \qquad [5.73]$$

$$P = f_{P2}\left(D, Q, N, g, \left(\theta_{in} - \theta_w\right), \mu_0, \gamma_0, \rho_0, c_{p0}, \lambda_0\right) \qquad [5.74]$$

Subsequently, the sign "*0*" linked to the material properties calculated at the reference temperature will be voluntarily omitted in order to simplify the presentation of the formulae.

Equations [5.73] and [5.74] use 11 physical quantities whose dimensions are expressed by four fundamental dimensions (L, M, T and K). Consequently, this problem is controlled by $11 - 4 = 7$ dimensionless numbers.

By writing the dimensional matrices associated with equations [5.73] and [5.74], and by transforming their core matrix into an identity matrix, it is possible to show that the spaces of the following dimensionless numbers associated with the two variables can be obtained by using as a base $\{\rho, N, d, \gamma\}$:

$$T_{out} = \gamma \cdot \left(\theta_{out} - \theta_w\right) = F_{\theta 1}\begin{cases} Re_N = \dfrac{\rho.N.D^2}{\mu}, \; Re_Q = \dfrac{\rho.Q}{\mu.D}, \; Fr = \dfrac{N^2.D}{g}, \\[2mm] Br = \dfrac{\gamma.\mu.N^2.D^2}{\lambda}, \; Pr = \dfrac{c_p.\mu}{\lambda}, \\[2mm] T_{in} = \gamma \cdot \left(\theta_{in} - \theta_w\right) \end{cases} \qquad [5.75]$$

$$Ne = \dfrac{P}{\rho.N^3.D^5} = F_P\begin{cases} Re_N = \dfrac{\rho.N.D^2}{\mu}, \; Re_Q = \dfrac{\rho.Q}{\mu.D}, \; Fr = \dfrac{N^2.D}{g}, \\[2mm] Br = \dfrac{\gamma.\mu.N^2.D^2}{\lambda}, \; Pr = \dfrac{c_p.\mu}{\lambda}, \\[2mm] T_{in} = \gamma \cdot \left(\theta_{in} - \theta_w\right) \end{cases} \qquad [5.76]$$

where the dimensionless numbers Re_N, Re_Q, Fr, Ne, Br and Pr correspond to the tangential and axial Reynolds numbers, and the Froude, Newton, Brinkman and Prandtl numbers, respectively. T_{out} and Ne represent the target internal measures of the problem. The six dimensionless numbers are responsible for the evolutions of the two target variables.

Rearrangements of certain dimensionless numbers are carried out so as to make the physical quantities D, Q, N and P appear separately in a single number, thus obtaining:

$$
\left\{
\begin{aligned}
\pi_D &= \left(\frac{Re_N^{\,2}}{Fr}\right)^{\!\!\frac{1}{3}} = D.\left[g.\left(\frac{\rho}{\mu}\right)^{\!2}\right]^{\!\frac{1}{3}} \\[2mm]
\pi_Q &= Re_Q.\left(\frac{Re_N^{\,2}}{Fr}\right)^{\!\!\frac{1}{3}} = Q.\left[g.\left(\frac{\rho}{\mu}\right)^{\!5}\right]^{\!\frac{1}{3}} \\[2mm]
\pi_N &= \left(\frac{Fr^2}{Re_N}\right)^{\!\!\frac{1}{3}} = N.\left[g^{-2}.\left(\frac{\mu}{\rho}\right)\right]^{\!\frac{1}{3}} \\[2mm]
\pi_P &= Ne.\left(Re_N^{\,7}.Fr\right)^{\!\frac{1}{3}} = P.\left[\left(\frac{\rho^4}{g.\mu^7}\right)\right]^{\!\frac{1}{3}}
\end{aligned}
\right.
$$

[5.77]

A rearrangement of the Brinkman number also helps to produce a purely material dimensionless number, denoted as Γ, whose variation depends solely on the physical properties of the material:

$$
\Gamma = \frac{Br}{\left(Re_N.Fr\right)^{\frac{2}{3}}.Pr^{\frac{5}{3}}} = \gamma.\left(\frac{g^2.\lambda^2}{\rho^2.c_p^{\,5}}\right)^{\!\frac{1}{3}}
$$

[5.78]

In the end, the configuration of the system is thus written as:

$$
T_{out} = F_{\theta 2}\left(\pi_D,\ \pi_Q,\ \pi_N,\ T_e,\ \Gamma,\ Pr\right)
$$

[5.79]

$$
\pi_P = F_{P2}\left(\pi_D,\ \pi_Q,\ \pi_N,\ T_e,\ \Gamma,\ Pr\right)
$$

[5.80]

The conditions of the problem, as specified at the beginning of the example, help to fix the numerical values of five of these numbers in the industrial prototype before any experiment is carried out:

$$\begin{cases} \pi_Q = 31.6 \\ T_{in} = 0.336 \\ T_{out} = 0.126 \\ \Gamma = 2.14 \times 10^{-9} \\ Pr = 4.55 \times 10^3 \end{cases} \qquad [5.81]$$

Subsequently, the problem can be summarized in order to establish the relationship between π_D and π_N so that T_{out} reaches the desired value. This can be written as follows:

$$\{\pi_D, \ \pi_N\} = 0 \qquad [5.82]$$

5.5.2.3. Extrapolation process

In order to comply with the conditions of similarity, and consequently, to guarantee that the phenomena observed on the scales of the model and the industrial equipment are identical (namely $T_{out} = T_{out}'$ and $\pi_P = \pi_P'$), it is necessary to maintain the same operating points on both scales, so that:

$$\begin{cases} \pi_D = \pi_D' \\ \pi_Q = \pi_Q' \\ \pi_N = \pi_N' \\ T_{in} = T_{in}' \\ \Gamma = \Gamma' \\ Pr = Pr' \end{cases} \qquad [5.83]$$

Following the convention used in previous guided examples, the dimensionless numbers expressed with and without primes refer to the scales of the model and the industrial prototype, respectively.

1) *Step 1*: definition of the model product and the reference temperature in the model.

The simultaneous equalities of the two purely material numbers, Γ and Pr, on both scales establish the characteristics of the fluid to use in the model. Consequently, that following conditions must be satisfied:

$$\begin{cases} \Gamma = \gamma.\left(\dfrac{g^2.\lambda^2}{\rho^2.c_p{}^5}\right)^{\!\frac{1}{3}} = \Gamma' = \gamma'.\left(\dfrac{g'^2.\lambda'^2}{\rho'^2.c'_p{}^5}\right)^{\!\frac{1}{3}} \\[4mm] Pr = \dfrac{c_p.\mu}{\lambda} = Pr' = \dfrac{c'_p.\mu'}{\lambda'} \end{cases}$$

[5.84]

Knowledge of the real fluid (the one used on the industrial scale) and the temperature taken as a reference in the prototype helps to fix the following quantities:

$$\theta_0 = 110°C, \ \mu(\theta_0), \ \rho(\theta_0), \ \lambda(\theta_0), \ c_p(\theta_0)$$

[5.85]

From here, an appropriate fluid may be selected and a new reference temperature can be chosen for the experiments on the model. To this end, Pawlowski [PAW 91] established, for various pure and mixed fluids, charts showing the variation of the dimensionless number Γ with the Prandtl number Pr, for different temperatures θ. It can thus be shown that it is possible to determine a fluid and define a reference temperature on the scale of the model, θ'_0 so that $Pr = Pr'$ and $\Gamma = \Gamma'$.

It is important to bear in mind that all the physical properties of real fluids and of those used in the model are defined at a reference temperature, θ_0 and θ'_0, respectively. Determining the model fluid and its properties $\mu', \rho', \lambda', c'_p$, is therefore done at a reference temperature θ'_0. We thus simultaneously define the characteristics of the model fluid and the reference temperature of the model:

$$\theta'_0, \mu'(\theta_0), \ \rho'(\theta_0), \ \lambda'(\theta_0), \ c'_p(\theta_0)$$

[5.86]

Pawlowski [PAW 91] specifies that the model fluid used in the model is a mixture of Baysilone oils (76% M10 and 24% M1000) at 25°C.

2) *Step 2*: definition of the size of the industrial equipment.

Knowledge of the physical properties of the fluid on the model helps us to calculate the diameter of the rotor in the industrial prototype, D, using the simultaneous equality on both scales:

$$\pi_D = \pi'_D$$

[5.87]

Since the diameter of the rotor in the model, D' is known, it is possible to deduce:

$$D = D' \cdot \left(\frac{\rho'/\mu'}{\rho/\mu} \right)^{2/3}$$

[5.88]

3) *Step 3*: definition of the temperature conditions.

Once the characteristics of the fluid and the reference temperature on the scale of the model are defined, it is possible to determine the temperature conditions to impose on this scale and thus define θ'_{in}, θ'_{out} and θ'_w using the system of three unknowns and the following three equations:

$$\begin{cases} \theta'_0 = \dfrac{\theta'_{in} + \theta'_w}{2} \\ \gamma' \cdot (\theta'_{in} - \theta'_w) = T'_{in} \\ \gamma' \cdot (\theta'_{out} - \theta'_w) = T'_{out} \end{cases}$$

[5.89]

Indeed, equality on both scales of the dimensionless numbers T_{in} and T_{out} imposes the conditions that:

$$\begin{cases} T'_{in} = T_{in} \\ T'_{out} = T_{out} \end{cases}$$

[5.90]

where the values of T_{in} and T_{out} are given by equation [5.81].

4) *Step 4*: definition of the volumetric flow rate Q'.

Knowledge of the volumetric flow rate Q in the prototype and the properties of the model fluid in the model help to deduce the value of the volumetric flow rate to be imposed on the model:

$$Q' = \pi_Q \cdot \left[g \cdot \left(\frac{\rho'}{\mu'} \right)^5 \right]^{-1/3}$$

[5.91]

where the value of π_Q is given in equation [5.81].

5) *Step 5*: experiments on the model.

The objective of this step is to define the rotation speeds N and the power P to apply to the industrial prototype.

The experiment thus consists of finding the rotation speed of the rotor in the model, N', at which the output temperature of product reaches θ'_{out}, and then of experimentally measuring the corresponding power P' to reach this operating point.

Then, the rotation speeds to impose N (respectively, the power P to apply) on the industrial scale are calculated using the simultaneous equality on both scales of the dimensionless numbers π_N and Ne:

$$\begin{cases} N = N'.\dfrac{(\mu'/\rho')^{\frac{1}{3}}}{(\mu/\rho)^{\frac{1}{3}}} \\[4mm] P = P'.\left(\dfrac{\rho'^4/\mu'^7}{\rho^4/\mu^7} \right)^{\frac{1}{3}} \end{cases} \qquad [5.92]$$

5.5.2.4. Conclusion

This guided example shows that the scale-up of a process involving a material with a variable physical property can be analyzed in the same way as for materials with constant properties. Nevertheless, it is necessary to add extra rules of similarity resulting from material similarity.

6

Case Studies

This chapter provides a collection of examples taken from our research in order to illustrate the application of dimensional analysis discussed in the previous chapters. These examples examine the study of momentum, mass and/or heat transfer and their consequences on a target variable of the process. They involve liquids, solids and gases in different types of isothermal and non-isothermal reactors. The dimensional analyses presented:

– are first "traditional" in their implementation (sections 6.1–6.3), with a series of examples on the rehydration of powders, the continuous foaming by whipping and the fouling of a plate heat exchanger;

– then, in the case of powder mixing (section 6.4), involve an intermediate variable as defined in Chapter 3, with a view to reduce the configuration of the system without restricting the range of validity of the process relationship;

– and finally, focus on processes involving fluids with variable physical properties (sections 6.5 and 6.6). These examples deal with the gas–liquid mass transfer in a mechanically stirred tank containing shear-thinning fluids and the ohmic heating of fluids; the variable property of the fluid is the apparent viscosity which depends on the shear rate and the electrical conductivity varying with temperature, respectively.

The notations and conventions associated with the design of the matrices used to generate internal measures are identical to those used in Chapter 2. In order to focus the user's attention on the key steps (listing of independent physical quantities, constructing and rearranging dimensionless numbers, and establishing and analyzing the process relationship), we decided not to include the details of the resolution methods used to carry out the change of base, nor the experimental programs.

However, we felt that it is important to describe and represent the configuration of the systems studied (star graph), reduced where possible, and the operating points explored (points or bars, see Chapter 2). The parameters (constant and exponents) involved in the process relationships (semi-empirical correlations) were systematically obtained by minimizing the sum of the squares of the differences between the values of the predicted target internal measures and their experimental values. The physical meaning and the validity of the process relationships obtained will be systematically discussed.

6.1. Rehydration time for milk powder in a stirred tank

Many liquid food products are concentrated and then dried in order to extend their shelf life by reducing the water activity. Before they are used, it is normally necessary to recombine the powder into liquid form. During this rehydration step, carried out in a mechanically stirred tank, the solid–liquid mass transfer can be described according to three stages:

– wetting/swelling which corresponds to the progressive penetration of water into the granular matrix;

– dispersion which corresponds to the fragmentation of the initial particles into subparticles;

– the final dissolution stage which corresponds to the disappearance of the dispersed particles.

As with the other operations involving solid–liquid mixtures, the scaling of this operation is closely connected to the interfacial area (mainly dependent on the powder granulometry) and to the overall coefficient of mass transfer K_l. For practical reasons relating to the expectations of the users, the target variable generally considered is the rehydration time t_r. It provides a general characterization of this mass transfer operation and gives a relatively accurate result of the time necessary so that certain functionalities of the reconstituted matter are reached.

The objective here is to use dimensional analysis to analyze how the rehydration of a powder in an aqueous mixture at a constant temperature, evaluated by t_r, is governed by the agitation process.

6.1.1. *Target variable and relevance list of independent physical quantities*

The rehydration time t_r, considered here as the *target variable*, corresponds to the time taken by the reconstituted mixture to reach certain functionalities starting from the moment when the powder is completely wetted by the fluid. The definition and the measure of this time are important and can be obtained, like mixing time, by using various methods. In this case, it was assessed through granulometric suspension analysis in volume, the rehydration time thus corresponding to the time taken for the powder to reach a given level of size reduction. An example of the change in average granulometric distribution in volume (given by the median diameter $d_{0.5}$) of a suspension of dairy powder (native[1] phosphocaseinate, also called micellar casein) during rehydration in stirred tank is shown in Figure 6.1.

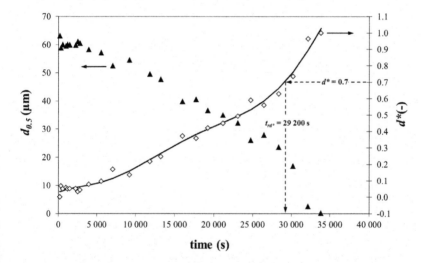

Figure 6.1. *Typical variation of the mean diameter $d_{0.5}$ and the reduction in size d* over time during rehydration of native phosphocaseinate in a stirred tank [JEA 10]*

---

1 *Native* refers to the original structure of the casein micelle which has been preserved in this case.

So that the mean diameters remain independent of the initial size of the powder and the final size to be achieved, the results are normalized by introducing the degree of rehydration, d^*, defined as:

$$d^* = \frac{d_p - d_{0.5}}{d_p - d_\infty}$$ [6.1]

where d_p and d_∞ represent the median diameters of the powder before initial wetting and at an infinite time, respectively. Here, $d_\infty \approx 200$ nm corresponds to the mean size of the casein micelles which is the major immiscible constituent of the product in this study. Thus, d^* is between 0 (zero rehydration) and 1 (complete rehydration). Subsequently, we will denote trd^* the rehydration time needed to reach a given degree of rehydration d^*. For instance, the rehydration time $tr0.7$ to reach a degree of rehydration d^* of 70% is equal to 29200 s (Figure 6.1).

The rehydration time t_{rd^*} is influenced by different independent physical quantities which can be classified in the following way (Figure 6.2):

– *boundary conditions*:

these are the geometric parameters which characterize the geometry of the tank (diameter T, height of the liquid H_L, etc.) and the impeller (diameter d, height of the blades w, clearance between the bottom of the tank and the tip of the impeller, C_b and a set of other geometric parameters denoted as $\{p_{geo}\}$);

– *material parameters*:

for dispersion, the solid–liquid mass fraction introduced ϕ is taken into account. In addition to the powder density ρ_p and to the initial mean diameter of the powder particles d_p, it is necessary to list the diffusion coefficient of the material D_m. Indeed, it can be assumed that convection alone does not explain the dissolution of the particles into the solution, especially in the hydrodynamic boundary layer close to the powder particle. Finally, the density ρ and the dynamic viscosity μ of the Newtonian liquid should be added;

– *process parameters*:

the rotation speed of the impeller is N;

– finally, the *universal constants*:

in this case, it is the gravitational acceleration *g*.

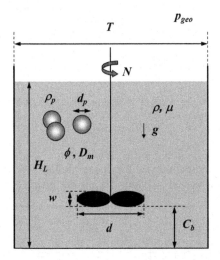

Figure 6.2. *Rehydration of milk powder: agitation system and associated parameters*

The problem can, therefore, be represented by the following dependence:

$$t_{rd*} = f\left(d, T, H_L, C_b, w, \{p_{geo}\}, \rho_p, D_m, \rho, \mu, \phi, d_p, N, g \right) \qquad [6.2]$$

6.1.2. Establishing dimensionless numbers

The physical quantities listed (equation [6.2]) are expressed in terms of fundamental dimensions in Figure 6.3. In this example, there are only three fundamental dimensions: mass, length and time.

According to the Vaschy–Buckingham theorem, 12 dimensionless numbers (15 physical quantities – 3 fundamental dimensions) should be determined to establish the process relationship describing the evolution of the rehydration time.

These *repeated variables* chosen to make up the base are ρ, N and d. Indeed, this base, in compliance with the principles set out in Chapter 2, is traditionally used in agitation problems. Therefore, it is logical to prioritize it over other options. The residual and core matrices are thus written as shown in Figure 6.3.

	t_{rd*}	d_p	ρ_p	μ	D_m	g	w	T	H_L	C_b	$\{p_{geo}\}$	ϕ	ρ	N	d
M	0	0	1	1	0	0	0	0	0	0	0	0	1	0	0
L	0	1	-3	-1	2	1	1	1	1	1	1	0	-3	0	1
T	1	0	0	-1	-1	-2	0	0	0	0	0	0	0	-1	0

Figure 6.3. *Rehydration time of milk powder in a stirred tank: dimensional matrix* **D** *(black outline), core matrix* **C** *(dark gray), and residual matrix* **R** *(light gray)*

Figure 6.4 shows the modified dimensional matrix obtained following the transformation of the core matrix into an identity matrix.

	t_{rd*}	d_p	ρ_p	μ	D_m	g	w	T	H_L	C_b	$\{p_{geo}\}$	ϕ	ρ	N	d
M	0	0	1	1	0	0	0	0	0	0	0	0	1	0	0
-T	-1	0	0	1	1	2	0	0	0	0	0	0	0	1	0
L+3M	0	1	0	2	2	1	1	1	1	1	1	0	0	0	1

Figure 6.4. *Rehydration time of milk powder in a stirred tank: modified dimensional matrix* **D**$_m$ *(black outline), identity matrix* **C**$_I$ *(dark gray), and modified residual matrix* **R**$_m$ *(light gray)*

The dimensionless numbers are formed using coefficients from the modified residual matrix. Therefore, the following set is obtained:

$$\left\{ \begin{array}{l} \pi_1 = N.t_{rd*} \ , \ \pi_2 = \dfrac{d_p}{d} \ , \ \pi_3 = \dfrac{\rho_p}{\rho}, \\[2mm] \pi_4 = \dfrac{\mu}{\rho.d^2.N} \ , \ \pi_5 = \dfrac{D_m}{d^2.N} \ , \ \pi_6 = \dfrac{g}{d.N^2}, \\[2mm] \pi_7 = \dfrac{T}{d} \ , \ \pi_8 = \dfrac{H_L}{d} \ , \ \pi_9 = \dfrac{w}{d} \ , \ \pi_{10} = \dfrac{C_b}{d} \ , \ \pi_{11} = \phi \ , \ \{\pi_{geo}\} \end{array} \right\}$$

[6.3]

It is interesting to note that:

– π_1 is analogous to a mixing number which is defined by the product of the rotation speed of the impeller and the mixing time (see section 6.4); π_1 will, therefore, be called the *rehydration number*, and will correspond to the number of rotations carried out by the impeller (proportional to the distance traveled by the impeller) in order to reach the degree of rehydration $d*$.

– dimensionless numbers π_4 and π_6 can be rearranged by raising them to the power of -1 in order to form Reynolds ($Re = \dfrac{\rho.N.d^2}{\mu}$) and Froude ($Fr = \dfrac{N^2.d}{g}$) numbers;

– the dimensionless number π_5 can be rearranged by dividing π_4 by π_5, in order to form the Schmidt number ($Sc = \dfrac{\mu}{\rho.D_m}$).

The set of dimensionless numbers associated with this process of rehydrating milk powder in a stirred tank is, therefore:

$$N.t_{rd^*} = F\left(\frac{d_p}{d}, \frac{\rho_p}{\rho}, Re,\ Sc,\ Fr, \frac{T}{d}, \frac{H_L}{d}, \frac{w}{d}, \frac{C_b}{d}, \phi, \left\{ \pi_{geo} \right\} \right) \qquad [6.4]$$

The configuration which characterizes the causes of the evolution in rehydration time is described by at least 11 internal measures, appearing in the second part of relationship [6.4].

6.1.3. *From the configuration of the system to the establishment of the process relationship*

Jeantet *et al.* [JEA 10] have performed rehydration tests at 30°C, in a thermostatic tank with a volume of 3.5 L, fitted with a crimped and inclined six-blade impeller to limit the formation of vortices. They have measured the evolution of the distribution of the particle size of a micellar casein powder as a function of time. During the experiments, various agitation conditions (6.67–16.67 rotations per second, rps) and mass fractions of powder ϕ (4.8–12% (w/w)) have been tested, while retaining:

– the agitation system;

– the type of powder (native miscellar casein), chosen for its poor capacity for rehydration, which therefore helps to more easily quantify the influence of different process conditions;

– the nature of the solvent (water).

Consequently, the form factors $\left(\dfrac{T}{d}, \dfrac{H_L}{d}, \dfrac{w}{d}, \dfrac{C_b}{d}, \{\pi_{geo}\} \right)$, the internal

measures $\dfrac{d_p}{d}$, $\dfrac{\rho_p}{\rho}$ and Sc remain constant. It will, therefore, not be possible to identify their influence on the rehydration number.

Finally, if there is no vortex formed around the impeller (no deformation of the free surface), it is then known that the Froude number has very little influence on the hydrodynamics within the tank.

This experimental program then suggests the simplified form of relationship [6.4]:

$$N.t_{rd^*} = F_2\left(Re, \phi\right) \qquad\qquad [6.5]$$

In these conditions and for the system being studied, only two internal measures, Re and ϕ, are responsible for the variation in the rehydration number. The graphical representation of this reduced configuration is shown in Figure 6.5.

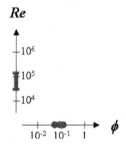

Figure 6.5. *Rehydration time of milk powder in a stirred tank: reduced configuration of the system*

The overall analysis of the results shows that a more precise rehydration time is obtained at a degree of rehydration of 70% ($d^* = 0.7$), which is far enough away from the end of the rehydration process. Consequently, the process relationship will be established for $t_{r0.7}$.

Figure 6.6 shows that for each solid–liquid fraction ϕ tested, the values of $N . t_{r0.7}$ do not depend on the Reynolds number Re. This product is, therefore, equal to a constant. This means that, in the range of Reynolds numbers studied, and for a given fraction ϕ, it is the number of rotations of the impeller which controls the degree of reduction in the size of the suspension. This result, which could have been predicted by the user simply on the basis of common sense, is demonstrated here via the application of dimensional analysis.

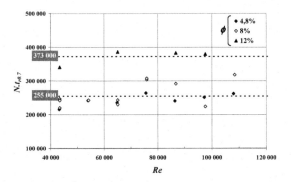

Figure 6.6. *Rehydration time of milk powder in a stirred tank: graphical representation of the process relationship [6.5] (number of turns made by the impeller to reach a reduction in powder size of 70%) [JEA 10]*

In this study, Jeantet *et al.* [JEA 10] showed that for $4.8\% \leq \phi \leq 8\%$, the values of $N \cdot t_{r0.7}$ varied little ($N \cdot t_{r0.7} = 255000$) irrespective of the value of *Re*, which could be explained by the high degree of homogeneity of the suspension over the range of impeller speeds. However, this is not the case for higher solid–liquid mass fractions (e.g., $\phi = 12\%$), where the value of $(N \cdot t_{r0.7}) = (373000)$ is higher than the preceding value.

These results are in agreement with other works in the literature. For instance, the Zwietering correlation [ZWI 58] helps to estimate the minimum rotation speed N_{min} for the suspension of powder particles in such a mixing system. This leads to a minimum speed N_{min} which increases with ϕ from 5.15 to 5.75 rps, and is therefore relatively close to the value for the minimum rotation speed used in the experimental program (6.67 rps). These values can explain the pronounced influence of the solid mass fraction on the rehydration time observed at $\phi = 12\%$.

With this knowledge, it is possible to easily predict how the rehydration time evolves with the rotation speed. Richard *et al.* [RIC 13] showed that, for a given solid mass fraction, the dependence of the rehydration number on the Reynolds number is a generic behavior which can be applied to other milk protein powders and types of impellers than those shown here. More generally, this case shows a strong analogy with the mixing time of a fluid into another fluid in a mechanically stirred tank. In this case, the number of impeller rotations to reach a desired degree of homogeneity is constant for a given flow regime, defined by a Reynolds number range.

6.2. Continuous foaming by whipping

In the food industry, foam products (e.g. milk or egg foam) are usually obtained by whipping in a continuous rotor-stator equipment. This process aims to introduce gas into a liquid in order to give specific texture properties to the product. These are directly linked to the two-phase structure produced, and in particular, to the size of the bubbles forming the foam and the quantity of gas incorporated. The challenge here is to understand the link between the size of the bubbles, the characteristics of the non-whipped base and the process parameters.

Owing to the complexity of the whipper geometry and the flow field generated, it is impossible to analytically resolve this problem and it is very time-consuming to carry out numerical simulations. In the literature, some attempts of modeling have been proposed using Taylor's work [TAY 32], via a specific capillary number:

$$Ca = \frac{\mu.\dot{\gamma}.d_b}{\sigma} \qquad [6.6]$$

where μ is the viscosity of a non-whipped base, $\dot{\gamma}$ is the mean shear rate, σ is the surface tension and d_b is the mean diameter of the bubbles. The limit of such an approach is that the shear rate is difficult to quantify, and this number does not indicate how the process parameters influence the mean shear rate [BAL 07].

In this example, we will carry out a dimensional analysis to overcome some of these difficulties, and identify the main internal measures which control this process.

6.2.1. Target variable and list of relevant independent physical quantities

The target variable chosen is the mean diameter of the bubbles at the whipping pressure, denoted by d_b. This target variable, which is preferred to the volume fraction of gas in the foam as traditionally used in whipping processes, is influenced by:

– *boundary conditions*:

anticipating the fact that a single machine was used in this study (Figure 6.7), only a characteristic length representative of the size of the equipment (and thus the scale of the process) was conserved in the relevance list of variables: the diameter of the rotor d. If several machines had been considered, it would have been necessary to add in the relevance list the other geometric parameters $\{p_{geo}\}$.

– *material parameters*:

- the parameters which characterize the liquid phase (dynamic viscosity μ, density ρ and surface tension σ);

- the parameters characterizing the gaseous phase (dynamic viscosity μ_g and density ρ_g).

– *process parameters*:

- the rotation speed of the rotor, N;

- the back pressure p which controls the expansion of the product at the outlet of the whipper;

- the volumetric flow rates of the liquid and gas, Q_L and Q_G.

– finally, *universal constants*:

- gravitational acceleration g.

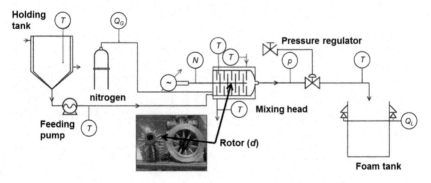

Figure 6.7. *Continuous foaming by whipping: representation of the whipping line, and photo of the disassembled mixing head (T: temperature, Q_G: volumetric flow rate of the gas, p: pressure, N: speed and Q_L: volumetric flow rate of the liquid) [MAR 13]*

The problem can, therefore, be written as follows:

$$d_b = f(d, \rho, \mu, \sigma, \rho_g, \mu_g, N, p, Q_L, Q_G, g) \qquad [6.7]$$

6.2.2. Establishing dimensionless numbers

The dimensional matrix shows that three fundamental dimensions are needed to express the dimensions of all the variables listed (Figure 6.8). Three variables are, therefore, necessary to make up the base.

The base chosen is $\{\rho, d, g\}$: indeed, as a single fluid and geometry are used in the experimental program, this choice enables us to obtain dimensionless numbers which only depend on one operating parameter. Another base could have been chosen, but rearrangements between the dimensionless numbers generated would have then been needed in order to obtain a comparable result. By carrying out linear combinations between different rows (Figure 6.9), the core matrix can be transformed into an identity matrix.

	d_b	μ	σ	ρ_g	μ_g	N	p	Q_L	Q_G	ρ	d	g
M	0	1	1	1	1	0	1	0	0	1	0	0
L	1	-1	0	-3	-1	0	-1	3	3	-3	1	1
T	0	-1	-2	0	-1	-1	-2	-1	-1	0	0	-2

Figure 6.8. *Dimensional matrix for studying continuous foaming by whipping*

0	d_b	μ	σ	ρ_g	μ_g	N	p	Q_L	Q_G	ρ	d	g
M	0	1	1	1	1	0	1	0	0	1	0	0
L+3M+0.5T	1	1.5	2	0	1.5	-0.5	1	2.5	2.5	0	1	0
-2T	0	0.5	1	0	0.5	0.5	1	0.5	0.5	0	0	1

Figure 6.9. *Modified dimensional matrix for studying continuous foaming by whipping*

The nine dimensionless numbers (12 physical quantities – 3 fundamental dimensions) of the second term of equation [6.8] are finally obtained:

$$\frac{d_b}{d} = F \left(\begin{array}{cccc} \dfrac{d^{0.5}.N}{g^{0.5}}, & \dfrac{\mu}{\rho.d^{1.5}.g^{0.5}}, & \dfrac{\rho_g}{\rho}, & \dfrac{\mu_g}{\rho.d^{1.5}.g^{0.5}}, \\[4mm] \dfrac{\sigma}{\rho.d^2.g}, & \dfrac{Q_L}{d^{2.5}.g^{0.5}}, & \dfrac{Q_G}{d^{2.5}.g^{0.5}}, & \dfrac{p}{\rho.d.g} \end{array} \right)$$

[6.8]

6.2.3. *From the configuration of the system to establishing a process relationship*

To successfully carry out this study, Mary *et al.* [MAR 13] created a pilot line in order to accurately measure the process parameters such as rotor speed, pressure and flow rate (since the geometry of the whipper is predefined). Model products were formulated so as to vary the levels of viscosity (0.4–1.4 Pa.s) and surface tension (40–49 N.m^{-1}) of the Newtonian fluids used. At the outlet of the whipper and at atmospheric pressure, the foams obtained were characterized in terms of the distribution of the size of the bubbles (measured by image analysis). Using this distribution, the mean diameter of the bubbles d_b (Sauter mean diameter) was calculated.

In the experimental program used by Mary *et al.* [MAR 13], a single gas was used. There was little variation between the different model fluids in terms of density ρ (around 1330 kg.m^{-3}) and it is, therefore, considered as constant. Consequently, the influence of the two internal measures $\dfrac{\rho_g}{\rho}$ and

$\dfrac{\mu_g}{\rho.d^{1.5}.g^{0.5}}$ cannot be studied.

The reduced configuration associated with this process is thus described by six internal measures grouped in the second term of equation [6.9][2]:

---

2 Given this space of internal measures, the user may be surprised by the relevance of the gravitational acceleration occurring in each internal measure, when its influence is presumably very small in a rotor/stator system. This result is simply due to the fact that g was chosen as a repeated variable. If rearrangements between these internal measures were carried out, it would be possible to only include g in one internal measure. Indeed, we will see subsequently that g has no significant influence on the target internal measure (equation [6.14]).

$$\frac{d_b}{d} = F\left(\frac{d^{0.5}.N}{g^{0.5}}, \frac{\mu}{\rho.d^{1.5}.g^{0.5}}, \frac{\sigma}{\rho.d^2.g}, \frac{Q_L}{d^{2.5}.g^{0.5}}, \frac{Q_G}{d^{2.5}.g^{0.5}}, \frac{p}{\rho.d.g}\right) \qquad [6.9]$$

The operating points explored are shown by the star graph in Figure 6.10.

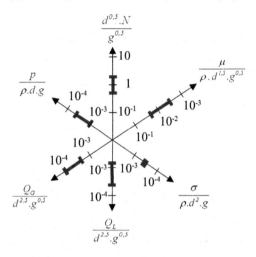

Figure 6.10. *Continuous foaming by whipping: reduced configuration of the system*

The adjustment of experimental results by monomial form leads to relationship [6.10]:

$$\frac{d_b}{d} = 109 \times 10^{-6}.\left(\frac{d^{0.5}.N}{g^{0.5}}\right)^{-0.45}.\left(\frac{\mu}{\rho.d^{1.5}.g^{0.5}}\right)^{-0.29}.\left(\frac{\sigma}{\rho.d^2.g}\right)^{0.16}.$$

$$\left(\frac{Q_L}{d^{2.5}.g^{0.5}}\right)^{0.09}.\left(\frac{Q_G}{d^{2.5}.g^{0.5}}\right)^{-0.04}.\left(\frac{p}{\rho.d.g}\right)^{-0.06} \qquad [6.10]$$

A comparison between the experimental diameters of the bubbles and those predicted in relationship [6.10] is shown in Figure 6.11.

It is interesting to note that:

– the last three internal measures in equation [6.10] (relating to the liquid flow rate, gas flow rate and pressure) have little influence over the diameter of the bubbles (weak exponents);

– on the contrary, the internal measures relating to viscosity, surface tension and rotation speed have a strong and significant impact.

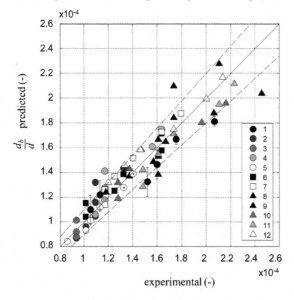

Figure 6.11. *Continuous foaming by whipping: comparison of the internal measures of the diameters of bubbles measured and calculated according to relationship [6.10]. The dotted lines correspond to a deviation of ± 10%*

Subsequently, model [6.10] can be reduced by only retaining these three internal measures:

$$\frac{d_b}{d} = b_0 \cdot \left(\frac{d^{0.5}.N}{g^{0.5}} \right)^{b_1} \cdot \left(\frac{\mu}{\rho.d^{1.5}.g^{0.5}} \right)^{b_2} \cdot \left(\frac{\sigma}{\rho.d^2.g} \right)^{b_3}$$ [6.11]

It is interesting to note that the first internal measure of the second term of equation [6.11] can be rearranged to include a capillary number Ca. This capillary number is the ratio of viscous forces (estimated at the periphery of the impeller) and the surface tension forces:

$$\frac{d^{0.5}.N}{g^{0.5}} \cdot \frac{\mu}{\rho.d^{1.5}.g^{0.5}} \cdot \frac{\rho.d^2.g}{\sigma} = \frac{\mu.N.d}{\sigma} = Ca$$ [6.12]

Finally, the following process relationship is obtained:

$$\frac{d_b}{d} = 503\times10^{-6}.Ca^{-0.46}.\left(\frac{\mu}{\rho.d^{1.5}.g^{0.5}}\right)^{0.11}.\left(\frac{\sigma}{\rho.d^2.g}\right)^{-0.05}$$ [6.13]

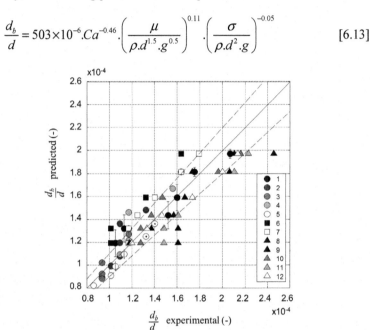

Figure 6.12. *Continuous foaming by whipping: comparison of internal measures of the diameters of bubbles measured and calculated according to relationship [6.13]. The dotted lines correspond to a deviation of ±10%*

Figure 6.12 shows that the model [6.13] gives a less faithful representation of the experimental results as the previous model [6.10]. This illustrates the small but real effect of the flow rates and pressure. Nevertheless, this new model shows the dominant influence of the capillary number on the size of the bubbles (exponent –0.46). Using this, a yet more simplified model can be obtained by further reducing the model as follows (Figure 6.13):

$$\frac{d_b}{d} = 372\times10^{-6}.Ca^{-0.4}$$ [6.14]

This relationship helps predict the mean size of the bubbles with less precision than that given by relationships [6.10] and [6.13]. However, it groups the influence of viscosity, rotor speed, rotor diameter and surface tension into a single dimensionless number: the capillary "process" number. Therefore, it indicates that the mechanism which governs the breakup of the bubbles is controlled by the equilibrium between shear forces and interfacial forces.

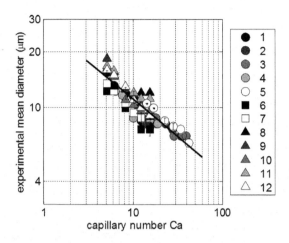

Figure 6.13. *Continuous foaming by whipping: representation of the diameters of bubbles measured according to the capillary number, as per relationship [6.14]*

6.3. Fouling of a plate heat exchanger by a milk protein solution

The heat treatment of solutions containing globular milk proteins helps to generate particles whose micrometric size controls the functional properties of these ingredients used in multiple food applications (e.g. fat replacers, etc.). This treatment leads to various mechanisms of protein denaturation, unfolding and aggregation [PET 11], which also causes the undesired adhesion of denatured proteins to the heated walls of the exchanger and its fouling. Controlling this limiting factor is crucial in order to manage:

– the quality of the products, due to the potential alteration of their nutritional qualities and the risk of developing micro-organisms which may be pathogenic within the deposits (biofilms);

– performance and energy cost due to the additional thermal resistance (deposits), the limited duration of production cycles and the necessary employment of long cleaning periods and negative environmental consequences [GRI 04, FRY 06].

Numerous parameters are involved in the fouling of heat exchanger walls by proteins, including [VIS 97, GRI 04, PET 11, PET 13]:

– the physical chemical characteristics and reactivity of the solution, protein and calcium concentration, pH, ionic strength, presence of dissolved gas and air bubbles, etc.;

– the heat treatment process: preheating system, heat profile programmed into the heat exchanger, temperature difference between the heat-transfer fluid and the product;

– the design and type of exchanger (material and geometry), which determine the properties of the exchange surface (surface energy and polarity) and influence the fluid flow of the fluid (wall shear stress).

Consequently, it is necessary to jointly characterize the impact of the process parameters on the thermal profile imposed and on the reactivity of the proteins in the product being processed in order to model the phenomenon of fouling in heat exchangers.

The reactivity of mainly milk-based globular protein, β-lactoglobulin, was characterized for various compositions (concentration of β-lactoglobulin and calcium) and in a wide range of temperatures. It can be described by a set of two successive kinetic reactions [PET 11]:

– unfolding, which represents the change from a native type (N) to an unfolded form (U);

– the aggregation of unfolded forms (irreversible formation of insoluble aggregates A).

At temperatures higher than 60°C (as per conditions in a heat exchanger), the denaturation mechanism of β-lactoglobulin can be written as follows:

$$N \xrightarrow{\ k_{un},n\ } U \xrightarrow{\ k_{ag},n\ } A \qquad\qquad\qquad [6.15]$$

$$-\frac{dN}{dt} = k_{un}.N^n \qquad\qquad\qquad [6.16]$$

$$\frac{dU}{dt} = k_{un}.N^n - k_{ag}.U^n \qquad\qquad\qquad [6.17]$$

$$\frac{dA}{dt} = k_{ag}.U^n \qquad\qquad\qquad [6.18]$$

where N, U and A represent the concentration of native, unfolded and aggregated form of β-lactoglobulin (in kg.m^{-3}), respectively, n is the order of the unfolding and aggregation reactions, and k_{un} and k_{ag} (in kg^{1-n}.m$^{-3(1-n)}$.s^{-1}) are the kinetic constants of the unfolding and aggregation reactions which depend on the temperature, as per the Arrhenius equation:

$$k_{un}(T) = k^{\circ}_{un} \exp\left(-\frac{E_{A,un}}{R.T}\right)$$
[6.19]

$$k_{ag}(T) = k^{\circ}_{ag} \exp\left(-\frac{E_{A,ag}}{R.T}\right)$$
[6.20]

where k°_{un} and k°_{ag} (in kg^{1-n}.m$^{-3(1-n)}$.s^{-1}) are the exponential Arrhenius factors of unfolding and aggregation of β-lactoglobulin, $E_{A,un}$ and $E_{A,ag}$ (J.mol^{-1}) are the activation energy of these reactions, R is the universal gas constant (8.314 J.mol^{-1}.K^{-1}) and T (K) is the temperature of the mixture.

The mechanism causing these denatured forms to create a deposit F on the heated walls of the exchanger is less well characterized, but can be considered as a kinetic reaction where the unfolded forms of β-lactoglobulin play a key role in the deposit [GRI 04, JUN 05, BLA 12, PET 13]. Consequently, a mathematical equation similar to [6.15]–[6.20] can be used to describe the fouling reaction:

$$U \xrightarrow{\quad k_{foul},n' \quad} F$$
[6.21]

$$\frac{dF}{dt} = k_{foul}.U^{n'}$$
[6.22]

$$k_{foul}(T) = k^{\circ}_{foul} \exp\left(-\frac{E_{A,foul}}{R.T}\right)$$
[6.23]

where F is the concentration of the fouling form of β-lactoglobulin, k_{foul} (kg$^{1-n'}$.m$^{-3(1-n')}$.s^{-1}) is the kinetic constant of the fouling reaction, n' (-) is the order of this reaction, k°_{foul} (kg$^{1-n'}$.m$^{-3(1-n')}$.s^{-1}) is the associated Arrhenius frequency factor and $E_{A,foul}$ (J.mol^{-1}) is the activation energy of the fouling reaction.

When the reactions of denaturation and fouling are considered simultaneously, the equations which describe the kinetic model of the reactivity of β-lactoglobulin correspond to equations [6.16], [6.18] and [6.22], to a modified from of equation [6.17]:

$$\frac{dU}{dt} = k_{un}.N^n - k_{ag}.U^n - k_{foul}.U^{n'} \qquad [6.24]$$

On the basis of the mechanisms proposed to describe the unfolding, aggregation and fouling of β-lactoglobulin, a "blind" dimensional analysis can be used in order to ascertain the impact of the heat transfer and reactivity on the size of the aggregates formed and the quantity of the deposit generated in a plate heat exchanger.

6.3.1. Target variables and list of relevant independent physical quantities

In order to simultaneously evaluate the process parameters of the heat treatment on the unfolding and aggregation of β-lactoglobulin for microparticulation and fouling on the plate heat exchanger, two *target variables* were defined:

– the mean diameter of β-lactoglobulin aggregates in volume at the outlet of the exchanger, d_{ag}, which especially affects their mouthfeel (smooth, floury or grainy texture);

– and the total mass of the protein deposit on all plates constituting the heat exchanger (measured after being oven-dried), M_{foul}.

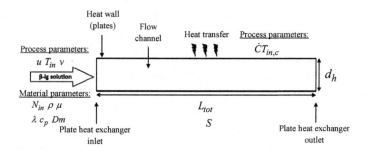

Figure 6.14. *Diagram of a plate heat exchanger [PET 13]. β-lg: β-lactoglobulin*

In this study, a single-channel plate heat exchanger is used. Its operation can be approximately described as a continuous reactor with a constant cross-section, in which the fluid flows with a constant superficial velocity, u, and receives a heat flux via a surface exchange S (Figure 6.14). This ignores the dead volume between the sections (which only represent a very small fraction of the total dead volume of the exchanger) and assumes a rectangular and straight geometry of equivalent hydraulic length and diameter.

The target variables relating to the denaturation and aggregation of the β-lactoglobulin and the fouling of the plate heat exchanger, d_{ag} and M_{foul}, are influenced by:

– *boundary conditions*: the equivalent length of the total path flowed in the exchanger, L_{tot}, the hydraulic diameter of the channel, d_h and the total surface of the walls enabling the heat transfer, S. This final parameter depends on L_{tot} and the width of the plates, which is a quantity which is not listed, and is independent from the other geometric quantities (d_h, L_{tot}). Therefore, it is necessary to include in the relevance list either the width of the plates or the surface S in addition to L_{tot} and d_h;

– *material parameters*: the physicochemical parameters characterizing the solution of β-lactoglobulin at the inlet of the exchanger – the mass concentration of native β-lactoglobulin, N_{in}, density, ρ, dynamic viscosity, μ, thermal conductivity, λ, the specific heat capacity, c_p and the molecular diffusivity of the native β-lactoglobulin, D_m;

– the kinetic parameters characterize the reactivity of β-lactoglobulin:

    i) order of unfolding and aggregation reactions, n;

    ii) order of fouling reaction, n';

    iii) pre-exponential Arrhenius factors and activation energies of the unfolding, aggregation and fouling reactions, $k°_{un}$, $k°_{ag}$, $k°_{foul}$ and $E_{A,un}$, $E_{A,ag}$, $E_{A,foul}$, respectively;

    iv) universal gas constant[3] R.

---

3 As specified in Chapter 3, the relevance list of physical quantities influencing the target variable must include universal physical constants, even though they act as constants. The ideal gas constant R is considered in this example where the chemical reactions are involved, like the universal gravitation constant g in the previous examples.

– *process parameters* characterizing the heat transfer:

- the inlet temperature of the product T_{in} which consequently constitutes the minimum temperature of the β-lactoglobulin solution in the exchanger;

- the superficial velocity of the β-lactoglobulin solution, u;

- the differences in temperature between the heat-transfer fluid and the β-lactoglobulin solution at the inlet, $T_{in,c} - T_{in}$, where $T_{in,c}$ represents the inlet temperature of the heat-transfer fluid;

- the ratio of heat capacity flow rates, $\dot{C}$, which is a dimensionless parameter characteristic of the heat exchange (equation [6.25]):

$$\dot{C} = \frac{\rho_c.c_{pc}.Q_c}{\rho.c_p.Q} \tag{6.25}$$

where ρ_c, c_{pc} and Q_c represent the density, the specific heat capacity and the mass flow rate of the heat-transfer fluid and Q represents the mass flow rate of the product (here, the β-lactoglobulin solution).

The total volume of the solution being treated, V, must also be listed, since the mass of the fouling deposited on the surface depends on the quantity of the product being treated. It would have also been possible to list the duration of the heat treatment instead of the total volume of the solution being treated, but this would have given the same result.

The problem can, therefore, be written as:

$$M_{foul} = f_1(L_{tot}, d_h, S, N_{in}, \rho, \mu, \lambda, c_p, D_m, n, n', k^{\circ}_{un}, k^{\circ}_{ag},$$
$$k^{\circ}_{foul}, E_{A,un}, E_{A,ag}, E_{A,foul}, R, T_{in}, T_{in,c} - T_{in}, u, \dot{C}, V) \tag{6.26}$$

$$d_{ag} = f_2(L_{tot}, d_h, S, N_{in}, \rho, \mu, \lambda, c_p, D_m, n, n', k^{\circ}_{un}, k^{\circ}_{ag},$$
$$k^{\circ}_{foul}, E_{A,un}, E_{A,ag}, E_{A,foul}, R, T_{in}, T_{in,c} - T_{in}, u, \dot{C}, V) \tag{6.27}$$

The number resulting from the independent physical quantities involved shows the complexity of this problem. We will see that dimensional analysis helps to significantly reduce the number of influencing to be taken into account.

6.3.2. *Establishing dimensionless numbers*

The dimensional matrix relating to the heat treatment of a β-lactoglobulin solution in a plate heat exchanger (Figure 6.15) is established by expressing the fundamental dimensions of the physical quantities involved (equations [6.26] and [6.27]). Five fundamental dimensions are involved in this problem (mass M, length L, time T, temperature K and quantity of matter N): the dimensionless base is, therefore, made up of five physical quantities. In this case $\{N_{in} ; k°_{un} ; R ; L_{tot} ; T_{in}\}$ have been chosen; these quantities do not vary in the experimental program, apart from variable T_{in}, although this can be considered as a reference temperature (the inlet temperature corresponding to the minimum temperature of the product in the exchanger).

The core matrix is then transformed into an identity matrix in order to obtain the following modified dimensional matrix (Figure 6.16).

The application of the Vaschy–Buckingham theorem with the base $\{N_{in} ; k°_{un} ; R ; L_{tot} ; T_{in}\}$ thus leads to a space of 19 dimensionless numbers (24 physical quantities – 5 fundamental dimensions) which help to characterize the fouling of a plate heat exchanger during the heat treatment of a β-lactoglobulin solution (equation [6.28]):

$$\frac{M_{foul}}{N_{in}.L_{tot}^{3}} =$$

$$F_1 \begin{pmatrix} \pi_1 = \dfrac{d_h}{L_{tot}}; \ \pi_2 = \dfrac{S}{L_{tot}^2}; \ \pi_3 = \dfrac{\rho}{N_{in}}; \ \pi_4 = \dfrac{\mu}{N_{in}^n.k°_{un}.L_{tot}^2}; \ \pi_5 = \dfrac{\lambda}{N_{in}^{3n-2}.k°_{un}^3.L_{tot}^4.T_{in}^{-1}}; \\[2ex] \pi_6 = \dfrac{c_p}{N_{in}^{2n-2}.k°_{un}^2.L_{tot}^2.T_{in}^{-1}}; \ \pi_7 = \dfrac{D_m}{N_{in}^{n-1}.k°_{un}.L_{tot}^2.T_e}; \ \pi_8 = n; \ \pi_9 = n'; \ \pi_{10} = \dfrac{k°_{ag}}{k°_{un}}; \\[2ex] \pi_{11} = \dfrac{k°_{foul}}{N_{in}^{n-n'}.k°_{un}}; \ \pi_{12} = \dfrac{E_{A,den}}{R.T_{in}}; \ \pi_{13} = \dfrac{E_{A,ag}}{R.T_{in}}; \ \pi_{14} = \dfrac{E_{A,foul}}{R.T_{in}}; \ \pi_{15} = \dfrac{T_{in,e} - T_{in}}{T_{in}}; \\[2ex] \pi_{16} = \dfrac{u}{N_{in}^{n-1}.k°_{un}.L_{tot}}; \ \pi_{17} = \dot{C}; \ \pi_{18} = \dfrac{V}{L_{tot}^3} \end{pmatrix} \quad [6.28]$$

	M_{foul}	d_h	S	ρ	μ	λ	c_p	D_m	n	n'	k°_{ag}	k°_{foul}	$E_{A,un}$	$E_{A,ag}$	$E_{A,foul}$	$T_{in,c}$-T_m	u	$\dot{C}$	V	N_m	k°_{um}	R	L_{tot}	T_{in}
M	1	0	0	-1	-1	1	0	0	0	0	1-n	1-n'	1	1	1	0	0	0	0	1	1-n	1	0	0
L	0	1	2	-3	-1	1	2	2	0	0	-3(1-n)	-3(1-n')	2	2	2	0	1	0	3	-3	-3(1-n)	2	1	0
T	0	0	0	0	-1	-3	-2	-1	0	0	-1	-1	-2	-2	-2	0	-1	0	0	0	-1	-2	0	0
K	0	0	0	0	0	-1	-1	0	0	0	0	0	0	0	1	1	0	0	0	0	0	1	0	1
N	0	0	0	0	0	0	0	0	0	0	0	0	-1	-1	-1	0	0	0	0	0	0	-1	0	0

Figure 6.15. Dimensional matrix for the fouling of a plate heat exchanger during the heat treatment of a β-lactoglobulin solution

	M_{foul}	d_h	S	ρ	μ	λ	c_p	D_m	n	n'	k°_{ag}	k°_{foul}	$E_{A,un}$	$E_{A,ag}$	$E_{A,foul}$	$T_{in,c}$-T_m	u	$\dot{C}$	V	N_{in}	k°_{un}	R	L_{tot}	T_{in}
$M + (1-n)T+(2n-1)N$	1	0	0	1	n	$3n-2$	$2n-2$	$n-1$	0	0	0	$n-n'$	0	0	0	0	$n-1$	0	0	1	0	0	0	0
$2N-T$	0	0	0	0	1	3	2	1	0	0	1	1	0	0	0	0	1	0	0	0	1	0	0	0
$-N$	0	0	0	0	0	0	0	0	0	0	0	0	1	1	1	0	0	0	0	0	0	1	0	0
$L+5N+3M$	3	1	2	0	2	4	2	2	0	0	0	0	0	0	0	0	1	0	3	0	0	0	0	0
$K-N$	0	0	0	0	0	-1	-1	0	0	0	0	0	1	1	1	1	0	0	0	0	0	0	0	1

Figure 6.16. Modified dimensional matrix for the fouling of a plate heat exchanger during the heat treatment of a β-lactoglobulin solution

The target internal measure $\dfrac{M_{foul}}{N_{in}.L_{tot}^{3}}$ compares the mass of the fouling with the mass of the heat-treated proteins, to within a multiplicative coefficient (equal to π_{18}). The dimensionless numbers in equation [6.28] can easily be rearranged to include conventional dimensionless numbers (Re, Pr, Sc, Da_I and Da_{III}) and material numbers characteristic of the reactivity of the β-lactoglobulin in the treated solution ($\dfrac{E_{A,ag}}{E_{A,un}}$ and $\dfrac{E_{A,foul}}{E_{A,un}}$):

– the Reynolds number, $Re = \dfrac{\rho \cdot u \cdot d_h}{\mu} = \dfrac{\pi_1.\pi_3.\pi_{16}}{\pi_4}$;

– the Prandtl number, $Pr = \dfrac{\mu \cdot c_p}{\lambda} = \dfrac{\pi_4.\pi_6}{\pi_5}$ and the Schmidt number,

$Sc = \dfrac{\mu}{\rho.D_m} = \dfrac{\pi_4}{\pi_3.\pi_7}$, which are two internal measures only involving physical quantities linked to the fluid;

– the first Damköhler number, $Da_I = \dfrac{L_{tot}}{u}.N_{in}^{n-1}.k^{\circ}_{un} = \dfrac{1}{\pi_{16}}$, which compares the mean residence time in the exchanger and the characteristic time of the unfolding reaction;

– the third Damköhler number enabling the evaluation of the impact of the exothermicity of the unfolding reaction of the β-lactoglobulin on the thermal profile of the exchanger, $Da_{III} = \dfrac{k^{\circ}_{un}.N_{in}^{n}.L_{tot}.\Delta H_{un}}{\rho.c_p.T_{in}.u} = \dfrac{\pi_{12}-1}{M_{\beta-lg}.\pi_3.\pi_6.\pi_{16}}$ where $\Delta H_{un} = E_{A,un} - R.T_{in}$ represents the mass activation energy of the unfolding reaction of the β-lactoglobulin and $M_{\beta-lg}$ is its molar mass ($\approx$ 18.3 kg.mol^{-1});

– $\dfrac{E_{A,ag}}{E_{A,un}} = \dfrac{\pi_{13}}{\pi_{12}}$;

– $\dfrac{E_{A,foul}}{E_{A,un}} = \dfrac{\pi_{14}}{\pi_{12}}$.

As a result, the space of dimensionless numbers relevant for the problem of fouling in a plate heat exchanger during the heat treatment of a β-lactoglobulin solution is as follows:

$$\frac{M_{foul}}{N_{in}.L_{tot}^{3}} = F_{1} \left(\begin{array}{c} \dfrac{d_{h}}{L_{tot}}; \dfrac{S}{L_{tot}^{2}}; \dfrac{\rho}{N_{in}}; Re; Pr; Da_{III}; Sc; n; n'; \dfrac{k^{\circ}_{ag}}{k^{\circ}_{un}}; \\[2mm] \dfrac{k^{\circ}_{foul}}{N_{in}^{n-n'}.k^{\circ}_{un}}; \dfrac{E_{A,un}}{R.T_{in}}; \dfrac{E_{A,ag}}{E_{A,un}}; \dfrac{E_{A,foul}}{E_{A,un}}; \dfrac{T_{in,c}-T_{in}}{T_{in}}; \\[2mm] Da_{I}; \dot{C}; \dfrac{V}{L_{tot}^{3}} \end{array} \right) \qquad [6.29]$$

Equation [6.29] shows that the configuration of the system has 18 internal measures. This configuration remains identical for the target internal measure linked to the denaturation and aggregation, $\dfrac{d_{ag}}{L_{tot}}$, since the reactions of denaturation, aggregation and fouling of the β-lactoglobulin are interdependent (equations [6.15]–[6.24]): $\dfrac{d_{ag}}{L_{tot}}$ enables the comparison between the size of the aggregates with a characteristic size of the plate heat exchanger (equation [6.30]).

$$\frac{d_{ag}}{L_{tot}} = F_{2} \left(\begin{array}{c} \dfrac{d_{h}}{L_{tot}}; \dfrac{S}{L_{tot}^{2}}; \dfrac{\rho}{N_{in}}; Re; Pr; Da_{III}; Sc; n; n'; \\[2mm] \dfrac{k^{\circ}_{ag}}{k^{\circ}_{un}}; \dfrac{k^{\circ}_{foul}}{N_{in}^{n-n'}.k^{\circ}_{un}}; \dfrac{E_{A,un}}{R.T_{in}}; \dfrac{E_{A,ag}}{E_{A,un}}; \dfrac{E_{A,foul}}{E_{A,un}}; \\[2mm] \dfrac{T_{in,c}-T_{in}}{T_{in}}; Da_{I}; \dot{C}; \dfrac{V}{L_{tot}^{3}} \end{array} \right) \qquad [6.30]$$

6.3.3. *From configurational analysis to the process relationship*

The experiments conducted by Petit *et al.* [PET 13] were carried out in a pilot plate heat exchanger. The heat treatment of the product was carried out in counter-current with the heat-transfer fluid (hot water) with one channel per pass: the β-lactoglobulin solution flowed in the exchanger in every other plate, alternating with the heat-transfer fluid. Four operating parameters varied independently during the experiment, determining the temperature profile to which the product in the exchanger was subjected (Figure 6.17):

– the inlet temperature of the product, T_{in};

– the residence time the product spends in the exchanger $t = \dfrac{L_{tot}}{u}$ corresponding to the time during which the temperature of the product is increased;

– the increase in product temperature between the inlet and outlet of the exchanger, $T_{out} - T_{in}$ (where T_{out} represents the product's outlet temperature), corresponding to the increase in temperature in the exchanger;

– the efficiency of the heat exchange toward the product, $\varepsilon = \dfrac{T_{out} - T_{in}}{T_{in,c} - T_{in}}$

(where $T_{in,c}$ represents the inlet temperature of the heat-transfer fluid), which determined the form of the temperature profile: convex for $\varepsilon < 90\%$, mainly linear for $\varepsilon = 90\%$ (reference process conditions), and concave for $\varepsilon > 90\%$.

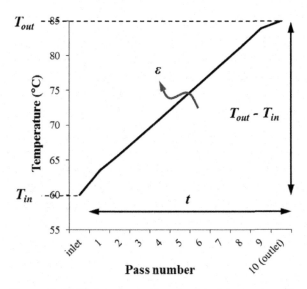

Figure 6.17. Link between the simulated thermal profile applied to the β-lactoglobulin solution in the plate heat exchanger (Sphere software, INRA Villeneuve d'Ascq) and operational parameters (reference conditions: $T_{in} = 60.3°C$; $T_{out} = 84.5°C$; $Q = 147$ L.h^{-1}; $T_{in,c} = 89.1°C$; $T_{out,c} = 63.9°C$; $Q_c = 159$ L.h^{-1})

The formation of aggregates in the β-lactoglobulin solution following heat treatment was tracked via the mean diameter in terms of the volume of aggregates, d_{ag}, measured by granulomorphometry [PET 12]. Moreover, the mass of the dry deposit on the plates of the exchanger M_{foul} was obtained by weighing the total of the deposits on each exchanger after being oven-dried [PET 13].

In the experimental program followed, the composition of the β-lactoglobulin solution, and therefore the kinetic parameters linked to its chemical reactivity, the volume of the solution treated and the geometry of the exchanger remained the same leading to a reduction in the configuration of the system as follows:

$$\left\{ \frac{\rho}{N_{in}};\ Re;\ Pr;\ Da_{III};\ Sc;\ \frac{E_{A,un}}{R.T_{in}};\ \frac{T_{in,c} - T_{in}}{T_{in}};\ Da_{I};\dot{C} \right\} \qquad [6.31]$$

It is necessary to reduce this configuration further since the experimental program does not allow certain internal measures to vary in a sufficiently wide range. Thus, $\frac{\rho}{N_{in}}$ must be removed because the density of the β-lactoglobulin solution at the inlet of the exchanger varies little (983 ± 6 kg.m^{-3}). Moreover, given the level of turbulence in the exchanger and the low level of thermal conductivity of the β-lactoglobulin solution, the heat transfer can be assumed to be dominated by convective phenomena, and the influence of the Prandtl number Pr can be considered as negligible. Likewise, the Schmidt number, Sc, can be excluded from this relationship because the turbulent flow of the β-lactoglobulin solution in the exchanger plays a key role in the mass transfer in relation to molecular diffusion. Finally, the heat flux produced by the chemical reactions of the β-lactoglobulin is negligible in comparison to the heat flux transferred from the heat-transfer fluid toward the product [PET 13]. Consequently, Da_{III} can also be removed from the process relationship.

The space of dimensionless numbers to take into account is eventually reduced to:

$$\left\{ Re;\ \frac{E_{A,un}}{R.T_{in}};\ \frac{T_{in,c} - T_{in}}{T_{in}};\ Da_{I};\dot{C} \right\} \qquad [6.32]$$

It can be noted that:

$-\dfrac{E_{A,un}}{R.T_{in}}$ and $\dfrac{T_{in,c}-T_{in}}{T_{in}}$ depend directly on T_{in}, while Re is indirectly and more weakly influenced by this temperature via the physicochemical properties of the solution (density and viscosity);

$-\dfrac{T_{in,c}-T_{in}}{T_{in}}$ and $\dot{C}$ are linked to T_{in}, $T_{out}-T_{in}$ and ε;

– Re and Da_I are both linked to the superficial velocity of the product and their respective influence on the target variables is difficult to determine. The choice was made to determine the process relationships with one of these dimensionless numbers, in order to identify which dominant phenomenon (flow regime or reactivity of the material) acts on the target variable. The one for the relationships providing the most accurate model on a case-by-case basis was retained.

At last, the process relationships required in this study have a monomial form for all the dimensionless numbers apart from $\dfrac{E_{A,un}}{R.T_{in}}$, for which an exponential dependence was chosen in order to include a form similar to the exponential Arrhenius factor[4], $\exp\left(-\dfrac{E_{A,un}}{R.T_{in}}\right)$, characteristic of the denaturation reaction of the β-lactoglobulin solution at the inlet of the exchanger (equations [6.19], [6.20] and [6.23]):

$$\frac{M_{foul}}{N_{in}.L_{tot}^{3}} = a_0.\left(Re \text{ or } Da_I\right)^{a_1}.\exp\left(a_2.\frac{E_{A,un}}{R.T_{in}}\right).\left(\frac{T_{in,c}-T_{in}}{T_{in}}\right)^{a_3}.\left(\dot{C}\right)^{a_4} \qquad [6.33]$$

$$\frac{d_{ag}}{L_{tot}} = b_0.\left(Re \text{ or } Da_I\right)^{b_1}.\exp\left(b_2.\frac{E_{A,un}}{R.T_{in}}\right).\left(\frac{T_{in,c}-T_{in}}{T_{in}}\right)^{b_3}.\left(\dot{C}\right)^{b_4} \qquad [6.34]$$

where a_i, b_i ($i = 0$-4) are the coefficients of the models.

---

4 As stated in Chapter 2, the monomial form is not necessarily the most suitable one for adjusting the "true" mechanistic law.

The reduced configuration of the system and the operating points explored are shown in Figure 6.18.

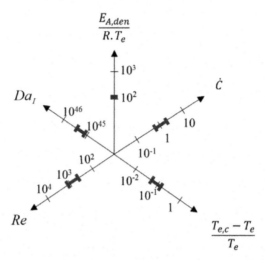

Figure 6.18. *Fouling of a plate heat exchanger during heating treatment of a milk protein solution: reduced configuration of the system*

The coefficients of models [6.33] and [6.34] were obtained using results from the experimental program. The models considering either Re or Da_I and which provide the best prediction of these results are shown below:

$$\frac{M_{foul}}{N_{in}.L_{tot}^{3}} = 7.4 \times 10^{-5}.(Da_{I})^{0.13}.\exp\left(-0.12 \times \frac{E_{A,un}}{R.T_{in}}\right).$$
$$\left(\frac{T_{in,c} - T_{in}}{T_{in}}\right)^{0.50}.\left(\dot{C}\right)^{-0.05}$$

[6.35]

$$\frac{d_{ag}}{L_{tot}} = 2.15 \times (Re)^{0.50}.\exp\left(-0.16 \times \frac{E_{A,un}}{R.T_{in}}\right).$$
$$\left(\frac{T_{in,c} - T_{in}}{T_{in}}\right)^{-0.46}.\left(\dot{C}\right)^{-1.25}$$

[6.36]

Figures 6.19 and 6.20 help highlight the accuracy of these models:

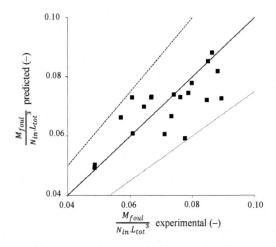

Figure 6.19. *Fouling of a plate heat exchanger during heat treatment of a milk protein solution: comparison of experimental and calculated internal measures of fouling according to relationship [6.35]. The dotted lines represent the margin for error $\pm 25\%$ ($R^2 = 44\%$)*

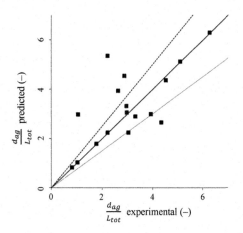

Figure 6.20. *Fouling of a plate heat exchanger during heat treatment of a milk protein solution: comparison of internal measures of aggregation measured and calculated according to relationship [6.36]. The dotted lines represent the margin for error $\pm 25\%$ ($R^2 = 33\%$)*

It can be seen in Figure 6.19 that the prediction of fouling mass over the entire plate heat exchanger using model [6.35] is correct. It is highly likely that a better characterization of the chemical reaction of the fouling of β-lactoglobulin (mechanism, Arrhenius prefactor and activation energy) and the inclusion of the kinetic parameters describing the reaction of fouling in the space of dimensionless numbers used could improve the accuracy of the model for predicting fouling mass.

However, the sizes of the aggregates are predicted with a lower level of accuracy (equation [6.36] and Figure 6.20, $R^2 = 33\%$), probably due to the poor repeatability of the analysis of aggregate size by granulomorphometry with low concentrations of β-lactoglobulin tested, as well as the slight variations in the composition of the solution.

When comparing models [6.35] and [6.36], it can be observed that Da_I allows for a better consideration of fouling by β-lactoglobulin, while Re is more suitable for describing aggregation. These two internal measures integrate the effect of the product flow rate, but Da_I enables the comparison of the characteristic time of reactions of β-lactoglobulin to the mean residence time in the exchanger, while Re is connected to the nature of the flow regime. This result shows that the formation of aggregates is more influenced by the flow regime than the formation of the fouling. Undoubtedly, this can be explained by the fact that aggregate formation takes place within the fluid volume and requires unfolded forms to be easily accessible, while the fouling takes place along the wall in a laminar boundary layer and is therefore, not significantly influenced by the flow regime within the body of the fluid.

Finally, these models clearly show the significant and combined influence of the reactivity of β-lactoglobulin and the temperature, $\dfrac{E_{A,un}}{R.T_{in}}$, during heat treatment. Indeed, an exponential form, recalling the exponential Arrhenius factor, helps to better describe its influence than a power function. To better ascertain the importance of this internal measure during the heat treatment processing in a plate heat exchanger, a reduced model with $\dfrac{E_{A,un}}{R.T_{in}}$ as the only variable was determined for the sole target internal measure $\dfrac{M_{foul}}{N_{in}.L_{tot}^{3}}$:

$$\frac{M_{foul}}{N_{in}.L_{tot}{}^3} = 8.94 \times 10^{-5} - \exp\left(-0.34.\left(144 - \frac{E_{A,un}}{R.T_{in}}\right)\right) \qquad [6.37]$$

Figure 6.21 compares the experimental values of the fouling mass to the predictions in relationship [6.37].

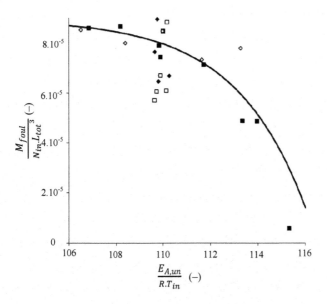

Figure 6.21. *Fouling of a plate heat exchanger during heating treatment of a milk protein solution: link between $M_{foul} / N_{in}.L_{tot}{}^3$ and $E_{A,un} / R.T_{in}$. Solid line: model [6.37]. Experimental results: solid squares: T_{in} variable; hollow squares: $T_{out} - T_{in}$ variable; solid diamonds: t variable; hollow diamonds: ε variable*

It can be observed that model [6.37] clearly takes account of the experimental values of the fouling mass, which confirms that the dimensionless numbers other than $\dfrac{E_{A,un}}{R.T_{in}}$ have a very moderate influence and that it is possible to correctly predict the fouling using a single process parameter (T_{in}) and knowledge of the kinetics of the β-lactoglobulin reaction. However, a similar approach applied to the size of the aggregates leads to a much higher level of dispersion, indicating that this target variable also depends significantly on other measures linked to the composition of the solution being treated.

6.4. Dry mixing of powders: mixing time and power consumption

The mixing of different powders is a crucial step in many food industry processes. It is often essential for obtaining a homogeneous mass to achieve the required properties and the conformity of the powder-based product to its specifications.

In this example, we suggest conducting a dimensional analysis in order to identify the dimensionless numbers which govern the mixing time and the power of a specific planetary mixer called Triaxe®, used here to mix powder-based foods. This mixer has two perpendicular revolution axes (Figure 6.22). First, this dimensional analysis will be conducted in the usual way. Afterwards, an intermediate variable, consisting of a characteristic velocity u_c, will be introduced into the relevance list of physical quantities instead of the rotation and gyration speeds (N_R and N_G). This is to reduce the configuration of the system without restricting the range of validity of the process relationship obtained.

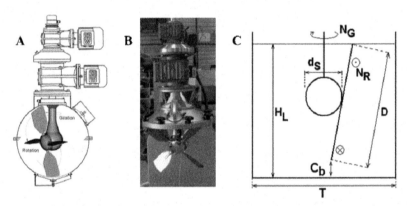

Figure 6.22. *Mixing of powders: Outline diagram A), photo B) and geometric and operational parameters C) used to define the planetary mixer Triaxe® [DEL 05]*

6.4.1. *List of relevant physical quantities*

In this case, we simultaneously take account of the two target variables which are the mixing time t_m and the power P. The physical quantities influencing the two target variables are:

– *boundary and initial conditions* which apply to the agitation system (Figure 6.22(c)): it is necessary to list the geometric parameters of the Triaxe® (T, D, d_s, C_b) and a set of other geometric parameters grouped into a set denoted by $\{p_{geo}\}$;

– *material parameters:* the flow domain contains two powders, characterized by their density (ρ_1 and ρ_2) and the mean diameter in terms of the volume of their particles (d_{p1} and d_{p2}). The mass fraction of the mixture $(x)^5$ and the total mass of the powder (m_p) are also listed;

– *process parameters*: these are the rotation and gyration speeds, N_R and N_G, of the mixer Triaxe®;

– *universal constants*: the gravitational acceleration g.

Thus, the problem can be written as follows:

$$P = f_1(T, D, d_s, C_b, \{p_{geo}\}, \rho_1, \rho_2, d_{p1},$$
$$d_{p2}, x, m_p, N_R, N_G, g) \tag{6.38}$$

$$t_m = f_2(T, D, d_s, C_b, \{p_{geo}\}, \rho_1, \rho_2, d_{p1},$$
$$d_{p2}, x, m_p, N_R, N_G, g) \tag{6.39}$$

6.4.2. *Establishing dimensionless numbers*

The dimensionless matrix associated with the physical quantities listed (equations [6.38] and [6.39]) is shown in Figure 6.23. For the same reasons as given previously (see section 6.1), the base chosen is made up of three repeated variables: $\{\rho_1, N_G, d_s\}$. For ease of writing, the two target variables have been placed in a single dimensional matrix. Normally, it would be necessary to write a separate dimensional matrix for each of the two target variables being studied.

---

5 If two phases are involved, it is not necessary to consider the mass fraction of each of the phases (in addition to the diameter of their particles and their density) because the sum of the mass fractions equals 1 and knowledge of one defines the other. By extension, the analysis of a multi-phase mixture of n different phases would require $3n$-1 material parameters to be considered (n diameter of particles, n density and n-1 mass fractions).

	P	t_m	ρ_2	d_{p1}	d_{p2}	x	m_p	N_R	g	T	D	C_b	$\{p_{geo}\}$	ρ_1	N_G	d_s
M	1	0	1	0	0	0	1	0	0	0	0	0	0	1	0	0
L	2	0	-3	1	1	0	0	0	1	1	1	1	1	-3	0	1
T	-3	1	0	0	0	0	0	-1	-2	0	0	0	0	0	-1	0

Figure 6.23. *Dimensional matrix for studying the power consumed and/or the mixing time of dry powders in a planetary mixer Triaxe[®]*

The transformation of the core matrix into an identity matrix helps to produce the modified forms of the residual and dimensional matrices, given in Figure 6.24.

	P	t_m	ρ_2	d_{p1}	d_{p2}	x	m_p	N_R	g	T	D	C_b	$\{p_{géo}\}$	ρ_1	N_G	d_s
M	1	0	1	0	0	0	1	0	0	0	0	0	0	1	0	0
-T	3	-1	0	0	0	0	0	1	2	0	0	0	0	0	1	0
L+3M	5	0	0	1	1	0	3	0	1	1	1	1	1	0	0	1

Figure 6.24. *Modified dimensional matrix for studying the power consumed and/or the mixing time of dry powders in a planetary mixer Triaxe[®]*

The modified dimensional matrix helps to produce the space of internal measures associated with the problem in the base $\{\rho_1 , N_G, d_s\}$:

$$N_{P_G} = \frac{P}{\rho_1 \cdot N_G^3 \cdot d_s^5} = F_1 \left(\begin{array}{c} Fr_G, \dfrac{\rho_2}{\rho_1}, \dfrac{d_{p1}}{d_s}, \dfrac{d_{p2}}{d_s}, x, \dfrac{m_p}{\rho_1.d_s^3}, \\ \dfrac{N_R}{N_G}, \dfrac{T}{d_s}, \dfrac{D}{d_s}, \dfrac{C_b}{d_s}, \{\pi_{geo}\} \end{array} \right)$$
[6.40]

and

$$\Theta_G = N_G.t_m = F_2 \left(\begin{array}{c} Fr_G, \dfrac{\rho_2}{\rho_1}, \dfrac{d_{p1}}{d_s}, \dfrac{d_{p2}}{d_s}, x, \dfrac{m_p}{\rho_1.d_s^3}, \\ \dfrac{N_R}{N_G}, \dfrac{T}{d_s}, \dfrac{D}{d_s}, \dfrac{C_b}{d_s}, \{\pi_{geo}\} \end{array} \right)$$
[6.41]

where $\{\pi_{geo}\}$, as before, represents the set of internal measures associated with the geometric parameters $\{p_{geo}\}$, and $Fr_G = \dfrac{N_G^{\,2}.d_s}{g}$ is a Froude number involving the gyration speed (in s^{-1}).

The configuration of the system therefore includes at least 11 internal measures, which are responsible for the variations of the two target variables, N_{pG} and Θ_G. These respectively correspond to a *power number* for a fixed speed ratio and a *mixing number* (number of gyrations necessary to reach a fixed degree of homogeneity for a fixed speed ratio). It is interesting to note that, compared to configuration obtained for traditional agitation systems (impeller rotating around a single vertically centered axis), the ratio of rotation and gyration speeds $\dfrac{N_R}{N_G}$, the Froude number and several additional shape factors are involved in this case. However, it is very logical that the Reynolds number, an internal measure of viscosity, does not appear in this relationship established for a granular mixture, unlike processes involving the mixing of miscible fluids.

6.4.3. From the configuration of the system to establishing the process relationship: the advantage of introducing an intermediate variable

The experimental program, as described by André *et al.* [AND 12], is based on a single mixture of powder and a single planetary mixer Triaxe®. It consists of mixing non-agglomerated semolina grains with a mean diameter d_{p1} equal to 340 µm with agglomerated semolina grains (called couscous semolina) with a mean diameter d_{p2} equal to 1400 µm. In order to monitor the progress of the mixing operation, the agglomerated particles of semolina were colored black. Therefore, it was possible to distinguish them from the light-colored non-agglomerated particles. Tests were carried out for a single mass fraction x: a constant mass of non-agglomerated semolina particles (7 kg) was mixed with a constant mass of agglomerated semolina grains (26.3 kg, or a total mass m_p of 33.3 kg) in fixed agitation conditions (N_R and N_G) over a given period of time, then stopping the mixer and pouring the contents onto a conveyor belt (Figure 6.25).

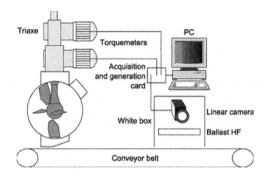

Figure 6.25. *Experimental device for mixing dry powders*
by a planetary mixer Triaxe® [AND 12]

The homogeneity of the mixture was assessed using a camera positioned above the conveyor belt. A threshold level was predefined for the images obtained in order to distinguish the agglomerated semolina particles (black pixels) from the non-agglomerated semolina particles (white pixels). The proportion of the black pixels was thus determined, then converted into a corresponding mass composition of the mixture (mass fraction of agglomerated semolina particles) using a pre-established calibration curve. It was then possible to establish the evolution of the variance of the mass fraction σ^2 depending on the time spent in the mixer (Figure 6.26), and to define a mixing index X_M according to the relationship:

$$X_M = \frac{\sigma_0^2 - \sigma^2}{\sigma_0^2 - \sigma_\infty^2} \qquad [6.42]$$

where σ_0^2 corresponds to the initial state (total segregation), defined by:

$$\sigma_0^2 = x.(1-x) \qquad [6.43]$$

σ_∞^2 is the asymptotic value reached by the variance. The mixing time t_m was defined as the time required to reach an X_M value of 0.99, corresponding to a variance σ^2 of 0.0021.

During these tests, the power consumed for mixing of the powders was monitored by a torquemeter.

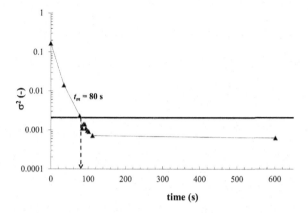

Figure 6.26. *Mixing of dry powders in a planetary mixer Triaxe®: determining the mixing time t_m [AND 12]*

With regard to this specific experimental program (agitation system and fixed powder characteristics), the configuration of the system (equations [6.40] and [6.41]) must be reduced as follows [AND 12]:

$$N_{P_G} = F_3 \left(Fr_G, \frac{N_R}{N_G} \right) \tag{6.44}$$

$$\Theta_G = F_4 \left(Fr_G, \frac{N_R}{N_G} \right) \tag{6.45}$$

The representation of this reduced configuration and the operating points explored is shown in Figure 6.27.

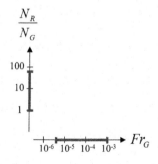

Figure 6.27. *Mixing of dry powders in a planetary system Triaxe®: reduced configuration of the system*

This approach represents the evolutions of the power number or the gyration mixing number depending on the Froude number, set by the value of the ratio $\dfrac{N_R}{N_G}$. In order to reduce the configuration of the system without restricting the range of validity of the established process relationship, it is also possible to introduce a characteristic speed u_c instead of the rotation and gyration speeds N_R and N_G. For a vertically centered impeller with a diameter d and a rotation speed N, the maximum speed which can be reached by an element of volume is the impeller tip speed, u_{tip} (rad.s^{-1}):

$$u_{tip} = \pi . N . d \qquad [6.46]$$

In this case, the maximum linear speed is independent of time. It can also be noted that for such an impeller, the speed usually considered to characterize the flow in a tank via the value of the Reynolds number is $N \cdot d$ ($Re = \dfrac{\rho \cdot (N \cdot d) \cdot d}{\mu} = \dfrac{\rho \cdot N \cdot d^2}{\mu}$). This shows that it is common not to consider the impeller tip speed $\pi \cdot N \cdot d$ as a characteristic speed, but this latter divided by π:

$$u_c = \frac{u_{tip}(t)}{\pi} \qquad [6.47]$$

For the Triaxe$^{\circledR}$ system which combines the dual motion (rotation and gyration), the instantaneous impeller tip speed $u_{tip}(t)$ is not a constant value with time in a fixed reference frame: indeed, it depends on the time and presents minimum and maximum values. The expression of the characteristic speed is therefore adapted as follows:

$$u_c = \frac{\max\left(u_{tip}(t)\right)}{\pi} \qquad [6.48]$$

The analytical expression of $u_{tip}(t)$ is obtained using, at each instant, the kinematics of a point located at the impeller tip (Figure 6.28). According to Delaplace et al. [DEL 05, DEL 07], it can be expressed as:

$$u_{tip}(t) = \sqrt{ \begin{aligned} &\left(\left(\frac{D}{2}\right) \cdot 2\pi \cdot N_R\right)^2 + \left(\left(\frac{d_s}{2}\right) \cdot 2\pi \cdot N_G\right)^2 - 2\pi^2 \cdot D \cdot d_s \cdot N_R \cdot \\ &N_G \cdot \sin\left(2\pi \cdot N_R \cdot t\right) + \left(2\pi \cdot N_G \cdot \left(\frac{D}{2}\right)\right)^2 \cdot \cos^2\left(2\pi \cdot N_R \cdot t\right) \end{aligned} } \qquad [6.49]$$

where t represents time.

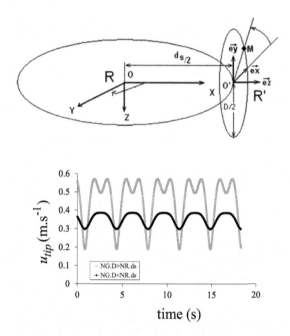

Figure 6.28. *Trajectory of a point M located at the tip of the blade and time variation of its linear speed* $u_{tip}(t)$ *for the agitation system Triaxe®️ according to various ratios* $N_R.d_s/(N_G.D)$ *[DEL 05, DEL 07].* R: *fixed reference frame;* R': *moving reference frame*

Depending on the value of the ratio $N_R \cdot d_s / (N_G \cdot D)$, the function $t \mapsto u_{tip}(t)$ does not show the same evolution over time (Figure 6.28). Indeed, it is characterized by the presence:

– of two extreme values (local minimum and maximum values) if $N_R \cdot d_s / (N_G \cdot D) \geq 1$;

– of four extreme values if $N_R \cdot d_s / (N_G \cdot D) < 1$.

The extreme values $\max(u_{tip}(t))$ are those obtained at the times for which the derivative of $u_p(t)$ (equation [6.49]) in relation to the time is canceled out. It can thus be shown that the characteristic speed u_c is equal to:

$$u_c = \sqrt{\left(N_R^2 + N_G^2\right)\cdot\left(d_s^2 + D^2\right)} \text{ when } N_R\cdot d_s/\left(N_G\cdot D\right) < 1 \qquad [6.50]$$

$$u_c = \left(N_R\cdot D + N_G\cdot d_s\right) \text{ when } N_R\cdot d_s/\left(N_G\cdot D\right) \geq 1 \qquad [6.51]$$

Since the analytical expression of u_c is known, this parameter can be introduced into the relevance list of physical quantities instead of the two individual rotational speeds (N_R; N_G). Depending on the chosen target variable (P or t_m), the problem can be written as:

$$P = f_3(T, D, d_s, C_b, \{p_{geo}\}, \rho_1, \rho_2, d_{p1}, d_{p2}, x, m_p, u_c, g) \qquad [6.52]$$

$$t_m = f_4(T, D, d_s, C_b, \{p_{geo}\}, \rho_1, \rho_2, d_{p1}, d_{p2}, x, m_p, u_c, g) \qquad [6.53]$$

The initial dimensional matrix (Figure 6.23) can therefore be rewritten by taking as a base the three repeated variables $\{\rho_1, u_c, d_s\}$ (Figure 6.29). The two target variables are once again written in a single dimensional matrix, for ease of writing.

	P	t_m	ρ_2	d_{p1}	d_{p2}	x	m_p	g	T	D	C_b	$\{p_{geo}\}$	ρ_1	u_c	d_s
M	1	0	1	0	0	0	1	0	0	0	0	0	1	0	0
L	2	0	-3	1	1	0	0	1	1	1	1	1	-3	1	1
T	-3	1	0	0	0	0	0	-2	0	0	0	0	0	-1	0

Figure 6.29. *Dimensional matrix for studying the power consumed or the mixing time of dry powders in a planetary mixer Triaxe® following the introduction of a characteristic speed u_c*

The transformation of the core matrix into an identity matrix helps to obtain the modified forms of the residual and dimensional matrices, shown in Figure 6.30.

	P	t_m	P_2	d_{p1}	d_{p2}	x	m_p	g	T	D	C_b	$\{p_{geo}\}$	ρ_1	u_c	d_s
M	1	0	1	0	0	0	1	0	0	0	0	0	1	0	0
-T	3	-1	0	0	0	0	0	2	0	0	0	0	0	1	0
L+3M+T	2	1	0	1	1	0	3	-1	1	1	1	1	0	0	1

Figure 6.30. *Modified dimensional matrix for studying the power consumed or the mixing time of dry powders in a planetary mixer Triaxe® following the introduction of a characteristic speed u_c*

Using the modified dimensional matrix, the dimensionless numbers can be determined. After some rearrangements of dimensionless numbers and the reduction of the configuration of the system due to the experimental program conducted, the following relationships are eventually obtained:

$$N_{Pm} = \frac{P}{\rho . u_c^3 . d_s^2} = F_5 \left(Fr_m = \frac{u_c^2}{g . d_s} \right) \qquad [6.54]$$

$$\Theta_m = \frac{u_c . t_m}{d_s} = F_6 \left(Fr_m \right) \qquad [6.55]$$

where N_{Pm}, Fr_m and Θ_m are the "modified" power, Froude and mixing numbers, respectively, in the sense that they include the characteristic speed u_c in their definition. The use of this intermediate variable consequently enables the reduction of the powder mixing problem to a relationship between two dimensionless numbers. The representation of this reduced configuration and the operating points explored are shown in Figure 6.31.

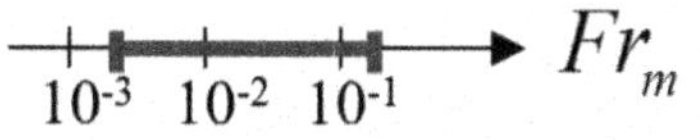

Figure 6.31. *Mixing of dry powders in a planetary mixer Triaxe® when a characteristic speed u_c is included: reduced configuration of the system*

6.4.4. Analysis of the process relationship obtained

Figures 6.32 and 6.33 show the evolution of the modified mixing and power numbers according to the modified Froude number. It can be noted that the introduction of a characteristic speed u_c offers the advantage of grouping all of the results onto a single mixing curve which characterizes the overall behaviour. The constant value of Θ_m shows that whatever the ratio of gyration and rotation speeds, it is the length of the path taken by the blade, represented by Θ_m, which enables us to obtain a homogeneous mixture of powder in a planetary mixer. This result is in line with those obtained for highly viscous liquids in traditional or planetary mixers in laminar regime [DEL 07]. Using the constant value of Θ_m, equal to 150, it is possible to deduce from equations [6.50] and [6.51] the expression of the powder mixing time t_m according to whether the ratio $N_R \cdot d_s / (N_G \cdot D)$ is lower than 1 or not:

$$t_m = \frac{150.d_s}{\sqrt{\left(N_R^2 + N_G^2\right).\left(d_s^2 + D^2\right)}} \quad \text{when } N_R \cdot d_s / (N_G \cdot D) < 1 \qquad [6.56]$$

$$t_m = \frac{150.d_s}{\left(N_R \cdot D + N_G \cdot d_s\right)} \quad \text{when } N_R \cdot d_s / (N_G \cdot D) \geq 1 \qquad [6.57]$$

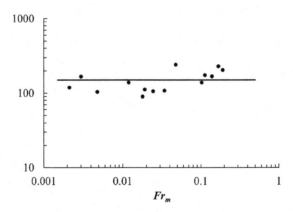

Figure 6.32. *Experimental mixing curve in the space* $\left(\Theta_m, Fr_m\right)$ *obtained by the planetary mixer Triaxe® during the mixing of dry powders [AND 12]*

In the same way, Figure 6.33 shows that it is possible to obtain a single master power curve when u_c is included instead of speeds N_R and N_G, whose equation can be approximated by:

$$N_{Pm} = \frac{15}{Fr_m}$$ [6.58]

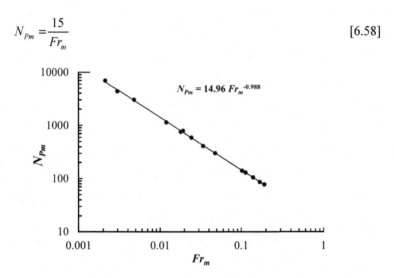

$$N_{Pm} = 14.96 \, Fr_m^{-0.988}$$

Figure 6.33. *Experimental power curve in the space* $\left(N_{Pm}, Fr_m \right)$ *obtained in the planetary mixer Triaxe® during the mixing of dry powders [AND 12]*

These results demonstrate that it is the characteristic speed u_c which governs the flow and the mixing of the powders. As a result, u_c is the key process parameter to take into account in order to control and optimize the mixing operation of dry powders in a planetary mixer.

6.5. Gas–liquid mass transfer in a mechanically stirred tank containing shear-thinning fluids

Mechanically stirred tanks are commonly used to carry out relatively fast chemical reactions or biological reactions (fermentation) which require the transfer of a constituent from a gaseous phase to a liquid phase. For instance, controlling the quality of wine depends highly on controlling the content of various types of dissolved gases, thus justifying the possible need for carbonization (injection of CO_2), decarbonization (injection of N_2) or oxygenation (injection of O_2). Likewise, the production or use of yeasts for transforming the contents of certain biological fluids (e.g. lactose in cheese making or starch in the glucose industry) requires the control of the gases dispersed within the bioreactor.

One of the key scaling parameters is the volumetric gas–liquid mass transfer coefficient, $k_l a$, defined as the product of the liquid-side mass transfer coefficient, k_l, and of the specific interfacial area, a. Even though there is an abundance of literature on this subject (e.g. [VAN 79, JUD 82, JOS 82, PAU 04, GAR 09]), the experimental data available and the associated models (dimensional or dimensionless correlations, mechanistic analysis and numerical simulations) mainly deal with Newtonian fluids with low or medium levels of viscosity. There are few studies however involving non-Newtonian fluids. In this case, the models suggested are limited to replacing the viscosity of Newtonian fluids by an apparent viscosity defined using the Ostwald-de-Waele rheological model and a mean shear rate deduced from the Metzner-Otto concept [MET 57]. Generally, it can be noted that the correlations obtained in this way do not give satisfactory results and/or cannot be readily generalized.

The aim of this example is to show that it is possible, using the tools presented in Chapter 4, to establish a robust process relationship which characterizes the factors controlling the volumetric gas–liquid mass transfer coefficient $(k_l a)$ when pseudoplastic fluids are aerated in a mechanically stirred tank. To do this, special attention must be given to the rheological model used to describe the shear-thinning behavior of the fluids involved.

6.5.1. *Case study with Newtonian fluids: target variable, list of relevant independent physical quantities and the configuration of the system*

In this study, the bubbles are directly generated in the liquid by a gas distributor placed underneath the impeller at the bottom of the tank. The gas–liquid mass transfer is therefore largely carried out through an aeration in volume. The contribution of the aeration by the free surface exists but is negligible. The volumetric gas–liquid mass transfer coefficient (k_la) is the *target variable* which is representative of the aeration conditions in the tank[6].

The physical quantities influencing k_la in the presence of Newtonian fluids are:

– *the boundary conditions* delineating the flow domain:

geometric parameters of the tank (diameter T, height of the fluid H_L), the impeller (diameter d, clearance between the bottom of the tank and the tip of the impeller C_b) and the gas distributor (diameter d_S). Likewise, we will note as $\{p_{geo}\}$ all of the geometric parameters, other than those previously defined, which are necessary for the complete description of the agitation and aeration system;

– *material parameters*:

these are viscosity and density of the gaseous and liquid phases ($\mu_g, \mu_l, \rho_g, \rho_l$), the surface tension of the liquid σ and the molecular diffusion coefficient of oxygen in the liquid phase D_m;

– *process parameters*:

these are the rotation speed of the impeller N, the superficial velocity of the gas u_g. The latter is an intermediate variable traditionally defined using the gas flow rate, Q_g, and the diameter of the tank, T, whereby:

$$u_g = \frac{Q_g}{\pi.T^2 / 4} \qquad \text{[6.59]}$$

– *universal constants*: the gravitational acceleration g.

---

6 In this example, we will not make a distinction between the overall liquid-side mass transfer coefficient, K_l, and the local liquid-side mass transfer coefficient, k_l, defined using the double film theory [LEW 24]. Indeed, the constituent which transfers from the gaseous phase to the liquid phase is oxygen and is not very soluble (very high Henry constant m). In this special case, we can show that, by applying the additivity rule of resistances to mass transfer: $1/K_l = 1/k_l + 1/(m.k_g) \approx 1/k_l$, where k_g is the local gas-side mass transfer coefficient.

We chose to list:

– quantities N and d, instead of power per unit volume (P/V) contrary to what is often encountered in the literature. The main reason is that P/V is an intermediate variable (as defined in Chapter 3), whose analytical expression is not known here;

– the superficial velocity of the gas, u_g, and not the flow rate of the gas, Q_g, because the parameter traditionally used in gas–liquid flows in a pipe or column (especially for establishing gas–liquid flow maps).

Therefore, in the presence of Newtonian fluids, the problem can be described as follows:

$$k_l a = f_N \left(u_g, N, d, T, d_S, H_L, C_b, \rho_l, \mu_l, \rho_g, \mu_g, \sigma, D_m, g, \{p_{geo}\} \right) \qquad [6.60]$$

where f_N represents the dimensional function linking the target variable ($k_l a$) to the influencing physical quantities in the case of Newtonian fluids (the index N refers to Newtonian fluids).

The dimensional matrix associated with equation [6.60] is shown in Figure 6.34. The repeated variables which were chosen to make up the base are: ρ_g, μ_g and g. This was in order to obtain dimensionless numbers depending on a minimum number of physical quantities which vary in the experimental program. The gas phase will indeed remain unchanged during the experiments – we will examine this in more detail later.

	$k_l a$	u_g	N	d	T	d_S	H_L	C_b	ρ_l	μ_l	σ	D_m	$\{p_{geo}\}$	ρ_g	μ_g	g
M	0	0	0	0	0	0	0		1	1	1	0	0	1	1	0
L	0	1	0	1	1	0	1	1	-3	-1	0	2	1	-3	-1	1
T	-1	-1	-1	0	0	1	0	0	0	-1	-2	-1	0	0	-1	-2

Figure 6.34. *Dimensional matrix for studying gas–liquid mass transfer in a stirred tank (Newtonian fluids)*

Figure 6.35 shows the modified dimensional matrix after the core matrix is transformed into an identity matrix.

	k_la	u_g	N	d	T	d_S	H_L	C_b	ρ_l	μ_l	σ	D_m	$\{p_{geo}\}$	ρ_g	μ_g	g
M+T+2A	1/3	-1/3	1/3	-2/3	-2/3	-2/3	-2/3	-2/3	1	0	-1/3	-1	-2/3	1	0	0
3M+L+T+A	-1/3	1/3	-1/3	2/3	2/3	2/3	2/3	2/3	0	1	4/3	1	2/3	0	1	0
A=-(3M+L+2T)/3	2/3	1/3	2/3	-1/3	-1/3	-1/3	-1/3	-1/3	0	0	1/3	0	-1/3	0	0	1

Figure 6.35. *Modified dimensional matrix for studying gas–liquid mass transfer in a stirred tank (Newtonian fluids)*

The dimensionless numbers are formed using the coefficients contained in the modified residual matrix (Figure 6.35). The following set is obtained in this way:

$$
\left\{
\begin{array}{ll}
\pi_1 = \dfrac{k_l a}{\rho_g^{1/3} \cdot \mu_g^{-1/3} \cdot g^{2/3}} \quad, & \pi_2 = \dfrac{u_g}{\rho_g^{-1/3} \cdot \mu_g^{1/3} \cdot g^{1/3}} \quad, \\[3mm]
\pi_3 = \dfrac{N}{\rho_g^{1/3} \cdot \mu_g^{-1/3} \cdot g^{2/3}} \quad, & \pi_4 = \dfrac{d}{\rho_g^{-2/3} \cdot \mu_g^{2/3} \cdot g^{-1/3}} \quad, \\[3mm]
\pi_5 = \dfrac{T}{\rho_g^{-2/3} \cdot \mu_g^{2/3} \cdot g^{-1/3}} \quad, & \pi_6 = \dfrac{d_S}{\rho_g^{-2/3} \cdot \mu_g^{2/3} \cdot g^{-1/3}} \quad, \\[3mm]
\pi_7 = \dfrac{H_L}{\rho_g^{-2/3} \cdot \mu_g^{2/3} \cdot g^{-1/3}} \quad, & \pi_8 = \dfrac{C_b}{\rho_g^{-2/3} \cdot \mu_g^{2/3} \cdot g^{-1/3}} \quad, \\[3mm]
\pi_9 = \dfrac{\rho_l}{\rho_g} \quad, & \pi_{10} = \dfrac{\mu_l}{\mu_g} \quad, \quad \pi_{11} = \dfrac{\sigma}{\rho_g^{-1/3} \cdot \mu_g^{4/3} \cdot g^{1/3}} \quad, \\[3mm]
\pi_{12} = \dfrac{D_m}{\rho_g^{-1} \cdot \mu_g} \quad, & \{\pi_{geo}\}
\end{array}
\right\}
\qquad [6.61]
$$

It is interesting to note that:

– π_1 (target internal measure) is identical to the dimensionless volumetric gas–liquid mass transfer coefficient proposed by Zlokarnik [ZLO 79];

– the product of π_4 with $(\pi_3)^2$ leads to the usual Froude number;

– the ratio π_{12} is the inverse of a modified Schmidt number, which is defined in terms of the properties of the gaseous phase (and not the liquid phase);

– shape factors characteristic of the geometry of the agitation and aeration system can be obtained by dividing each of the numbers π_5, π_6, π_7 and π_8 by π_4.

Following these rearrangements and after introducing the kinematic viscosity of the gas ($v_g = \mu_g/\rho_g$), equation [6.61] becomes:

$$k_l a^* = k_l a. \left(\frac{v_g}{g^2} \right)^{1/3} = F_N \begin{pmatrix} U_g^* = \dfrac{u_g}{(v_g.g)^{1/3}} \ , \ Fr = \dfrac{N^2.d}{g} \ , \\[2mm] d^* = d. \left(\dfrac{g}{v_g^2} \right)^{1/3} \ , \ \dfrac{T}{d} \ , \ \dfrac{d_S}{d} \ , \\[2mm] \dfrac{H_L}{d} \ , \ \dfrac{C_b}{d} \ , \ \rho^* = \dfrac{\rho_l}{\rho_g} \ , \ \mu^* = \dfrac{\mu_l}{\mu_g} \ , \\[2mm] \sigma^* = \dfrac{\sigma}{(\rho_g^3.v_g^4.g)^{1/3}} \ , \ Sc_g = \dfrac{v_g}{D_m} \\[2mm] \{\pi_{geo}\} \end{pmatrix} \qquad [6.62]$$

where $\{\pi_{geo}\}$ represents the set of geometric internal measures, other than $\dfrac{T}{d}$, $\dfrac{d_S}{d}$, $\dfrac{C_b}{d}$, $\dfrac{H_L}{d}$, coming from $\{p_{geo}\}$, and F_N the process relationship required in the case of Newtonian fluids.

Therefore, dimensional analysis suggests that the target internal measure (the dimensionless volumetric gas–liquid mass transfer coefficient, $k_l a^*$) is potentially affected by at least 12 dimensionless numbers present in the second part of equation [6.62], which are internal measures of the superficial velocity of the gas (U_g^*) and of the deformation of the free surface due to gravity and rotation speed (Fr), geometric internal measures (d^*, $\dfrac{T}{d}$, $\dfrac{d_S}{d}$, $\dfrac{C_b}{d}$, $\dfrac{H_L}{d}$, $\{\pi_{geo}\}$), internal measures of density (ρ^*), viscosity (μ^*), surface tension (σ^*) and molecular diffusion (Sc_g). These numbers represent the configuration of the system, which is the complete set of the internal measures which characterize the causes of the evolution of $k_l a$.

6.5.2. *Extension to shear-thinning fluids*

In this case, the variable physical property of the material is the apparent viscosity μ_a of the liquid phase which depends on the shear rate $\dot{\gamma}$: $s(p) = \mu_a(\dot{\gamma})$. In this study, two models were tested to describe the rheological behavior of the shear-thinning fluids:

– the Ostwald-de-Waele model (power law):

$$\mu_a(\dot{\gamma}) = k.\dot{\gamma}^{n_{ost}-1} \qquad [6.63]$$

where k and n_{ost} are the consistency and the flow index of the fluid,[7] respectively;

– the Williamson–Cross model:

$$\mu_a(\dot{\gamma}) = \frac{\mu_w}{1 + (t_w.\dot{\gamma})^{1-n_w}} \qquad [6.64]$$

where μ_w is a parameter which represents the viscosity of the fluid for the low shear rates where it behaves as a Newtonian fluid, t_w is a characteristic time related to the transition between Newtonian and purely pseudoplastic behaviors and n_w is the flow index.

According to the rules given in Chapter 4, the set of internal measures characteristic of the material $\{\pi_m\}$ to add to the initial relevance list established for Newtonian fluids (equation [6.60]) depends on the rheological model used:

– for the Ostwald-de-Waele model, they are:

$$\{\pi_m\} = \{n_{ost}\} \qquad [6.65]$$

– and, for the Williamson–Cross model, they are:

$$\{\pi_m\} = \{n_w, t_w.\dot{\gamma}_0\} \qquad [6.66]$$

---

7 The flow index associated with the Ostwald-de-Waele model is denoted as n_{ost} in order to differentiate it from the flow index relating to the Williamson–Cross model, denoted as n_w.

We have seen that:

– the material function associated with the Ostwald-de-Waele model is reference-invariant (see Chapter 4, section 4.3.4). Consequently, the reference shear rate $\dot{\gamma}_0$ must not be added to the initial relevance list. However, a value of $\dot{\gamma}_0$ must be chosen in order to calculate μ_0: we arbitrarily chose, $\dot{\gamma}_0 = N$ meaning that:

$$\mu_0 = \mu_a(N) = k.N^{n_{ost}-1} \qquad [6.67]$$

– the material function associated with the Williamson–Cross model is not reference-invariant (see Chapter 4, section 4.3.4), so the reference shear rate $\dot{\gamma}_0$ must be listed. We have seen (see Chapter 4, section 4.4.3) that it is relevant to consider it as being equal to $1/t_w$ to reduce the number of internal measures characteristic of the material (indeed, in this case, the dimensionless number $t_w \cdot \dot{\gamma}_0$ becomes equal to 1, whatever fluid is used). It should be recalled that in these conditions the apparent viscosity calculated as $\dot{\gamma}_0 = \dfrac{1}{t_w}$, μ_0, is invariably:

$$\mu_0 = \mu_a\left(\frac{1}{t_w}\right) = \frac{\mu_w}{1+\left(t_w \cdot \dot{\gamma}_0\right)^{1-n}} = \frac{\mu_w}{1+\left(t_w \cdot \dfrac{1}{t_w}\right)^{1-n}} = \frac{\mu_w}{2} \qquad [6.68]$$

Thus, the problem can be expressed in terms of physical quantities influencing the volumetric gas–liquid mass transfer coefficient:

– when the rheological behavior of fluids is described by the Ostwald-de-Waele model, by:

$$k_l a = f_{ost}(u_g, N, d, T, d_S, H_{L,}, C_b, \rho_l, \rho_g, \mu_g, \sigma, D_m, g, \{p_{geo}\},$$
$$\mu_0, n_{ost}) \qquad [6.69]$$

where f_{ost} represents the dimensional function linking the target variable ($k_l a$) to the influencing physical quantities, μ_0 is calculated using equation [6.67];

– when the rheological behavior of the fluids is described by the Williamson–Cross model, by:

$$k_l a = f_{wc}(u_g, N, d, T, d_S, H_L, C_b, \rho_l, \rho_g, \mu_g, \sigma, D_m, g, \{p_{geo}\},$$

$$\dot{\gamma}_0 = \frac{1}{t_w}, \mu_0, n_w)$$

[6.70]

where f_{wc} represents the dimensional function linking the target variable ($k_l a$) to influencing physical quantities, μ_0 is calculated using equation [6.68].

As a result, it is necessary to add, in the dimensional matrix written for the Newtonian fluids (Figure 6.34), the quantities μ_0 and n_{ost} in the case of the Ostwald-de-Waele model, and the quantities $\dot{\gamma}_0 = \frac{1}{t_w}$, μ_0 and n_w for the case of the Williamson–Cross model respectively. By choosing the same set of repeated variables as in the Newtonian case, the modified dimensional matrices are obtained (Figures 6.36 and 6.37).

$k_l a$	u_g	N	d	T	d_S	H_L	C_b	ρ_l	σ	D_m	$\{p_{geo}\}$	μ_0	n_{ost}	ρ_g	μ_g	g
1/3	-1/3	1/3	-2/3	-2/3	-2/3	-2/3	-2/3	1	-1/3	-1	-2/3	0	0	1	0	0
-1/3	1/3	-1/3	2/3	2/3	2/3	2/3	2/3	0	4/3	1	2/3	1	0	0	1	0
2/3	1/3	2/3	-1/3	-1/3	-1/3	-1/3	-1/3	0	1/3	0	-1/3	0	0	0	0	1

Figure 6.36. *Modified dimensional matrix for studying the gas–liquid mass transfer in a stirred tank when the rheological behavior of the fluids is described by the Ostwald-de-Waele model*

$k_l a$	u_g	N	d	T	d_S	H_L	C_b	ρ_l	σ	D_m	$\{p_{géo}\}$	$\dot{\gamma}_0$	μ_0	n_w	ρ_g	μ_g	g
1/3	-1/3	1/3	-2/3	-2/3	-2/3	-2/3	-2/3	1	-1/3	-1	-2/3	1/3	0	0	1	0	0
-1/3	1/3	-1/3	2/3	2/3	2/3	2/3	2/3	0	4/3	1	2/3	-1/3	1	0	0	1	0
2/3	1/3	2/3	-1/3	-1/3	-1/3	-1/3	-1/3	0	1/3	0	-1/3	2/3	0	0	0	0	1

Figure 6.37. *Modified dimensional matrix for studying the gas–liquid mass transfer in a stirred tank when the rheological behavior of the fluids is described by the Williamson–Cross model*

By using the coefficients contained in the modified residual matrices given in Figures 6.36 and 6.37, and carrying out rearrangements between dimensionless numbers, the following sets of dimensionless numbers can be obtained:

– for the Ostwald-de-Waele model:

$$
k_l a^* = k_l a . \left(\frac{\upsilon_g}{g^2} \right)^{1/3} = F_{ost}
\begin{cases}
U_g^* = \dfrac{u_g}{(\upsilon_g \cdot g)^{1/3}} \ , \ Fr = \dfrac{N^2 . d}{g} \ , \\[2ex]
d^* = d . \left(\dfrac{g}{\upsilon_g^2} \right)^{1/3} \ , \ \dfrac{T}{d} \ , \ \dfrac{d_s}{d} \ , \\[2ex]
\dfrac{H_L}{d} \ , \ \dfrac{C_b}{d} \ , \ \rho^* = \dfrac{\rho_l}{\rho_g} \ , \ \mu^* = \dfrac{\mu_0}{\mu_g} \ , \\[2ex]
\sigma^* = \dfrac{\sigma}{(\rho_g^3 . \upsilon_g^4 . g)^{1/3}} \ , \ Sc_g = \dfrac{\upsilon_g}{D_m} \\[2ex]
\{ \pi_{geo} \} \ , \ n_{ost}
\end{cases}
\qquad [6.71]
$$

where F_{ost} is the process relationship required;

– and for the Williamson–Cross model:

$$
k_l a^* = k_l a . \left(\frac{\upsilon_g}{g^2} \right)^{1/3} = F_{wc}
\begin{cases}
U_g^* = \dfrac{u_g}{(\upsilon_g \cdot g)^{1/3}} \ , \ Fr = \dfrac{N^2 . d}{g} \ , \\[2ex]
d^* = d . \left(\dfrac{g}{\upsilon_g^2} \right)^{1/3} \ , \ \dfrac{T}{d} \ , \ \dfrac{d_s}{d} \ , \\[2ex]
\dfrac{H_L}{d} \ , \ \dfrac{C_b}{d} \ , \ \rho^* = \dfrac{\rho_l}{\rho_g} \ , \ \mu^* = \dfrac{\mu_0}{\mu_g} \ , \\[2ex]
\sigma^* = \dfrac{\sigma}{(\rho_g^3 . \upsilon_g^4 . g)^{1/3}} \ , \ Sc_g = \dfrac{\upsilon_g}{D_m} \\[2ex]
\{ \pi_{geo} \} \ , \ t_w^* = \dfrac{1}{t_w} \left(\dfrac{\upsilon_g}{g^2} \right)^{1/3} \ , \ n_w
\end{cases}
\qquad [6.72]
$$

where F_{wc} is the process relationship required. It should be noted that, even if the dimensionless number $(t_w \cdot \dot{\gamma}_0)$ no longer appears explicitly in equation [6.72] due to the relevant choice of the reference shear rate ($\dot{\gamma}_0 = \dfrac{1}{t_w}$), it is still part of the configuration of the system, but its value remains equal to 1.

Compared to Newtonian fluids, the configuration of the system is therefore increased by one and two internal measures (n_{ost} and $\{t_w^*, n_w\}$, respectively) when the rheological behavior of the fluid is described by the Ostwald-de-Waele and the Williamson–Cross models, respectively.

6.5.3. From the configuration of the system to establishing the process relationship

In the experimental program used to establish the process relationship [HAS 12], the gaseous phase (air), and the geometry and the dimensions of the tank, the impeller and the gas distributor remained unchanged. A graphical representation of the experimental device is shown in Figure 6.38.

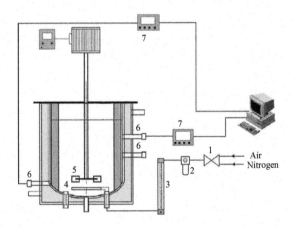

Figure 6.38. *Study of gas–liquid mass transfer in a stirred tank: experimental device (1: valve; 2: monometer; 3: flowmeter; 4: air distributor; 5: Chemineer® impeller; 6: temperature and dissolved oxygen probes; 7: oxygen transmitter)*

The gas flow rates varied from 0.33 to 3.33 L.min^{-1}, and the rotation speed of the impeller from 200 to 1000 rpm.

Seven liquid phases were tested: three Newtonian fluids (water, aqueous solutions of glycerol) and four shear-thinning fluids (aqueous solutions of CMC and xanthan gum). Their rheological behavior was measured and the parameters of each rheological model (k and n_{ost} for the Ostwald-de-Waele model, μ_w, t_w and n_w for the Williamson–Cross model) were determined. It was verified that these fluids had no viscoelastic properties.

The volumetric gas–liquid mass transfer coefficient ($k_l a$) was measured by two methods:

– a physical method, for which two dissolved oxygen probes located at the bottom and the top of the tank were used (Figure 6.38);

– the chemical method developed by Painmanakul et al. [PAI 05].

The small contribution of the aeration by the free surface was also verified. This experimental program prompts us to reduce the configuration of the system as established in equations [6.62], [6.71] and [6.72]. Indeed:

– a single device (tank, impeller and air distributor) was tested. The influence of the geometric internal measures on $k_l a^*$ cannot therefore be evaluated;

– since the value of the diffusion coefficients of oxygen in the viscous fluids used is not known, is it assumed to be equal to that of water at 20°C ($D_m = 2.10^{-9}$ m^2.s^{-1}). As a result, the process relationship will only be valid for a modified Schmidt number Sc_g close to 7850, corresponding to this particular value of D_m.

– the density of the various fluids tested varies slightly ($847 < \rho^* < 1015$). The contribution of this internal measure will therefore be disregarded, even though it is highly likely that its influence will remain low.

Consequently, the configuration of the system must be reduced to:

– four internal measures for Newtonian fluids:

$$k_l a^* = F_{N,r}\left(U_g^*, Fr, \mu^*, \sigma^* \right) \qquad [6.73]$$

– five internal measures when the rheological behavior of the fluids is described by the Ostwald-de-Waele model:

$$k_l a^* = F_{ost,r}\left(U_g^*, Fr, \mu^*, \sigma^*, n_{ost} \right) \qquad [6.74]$$

– six internal measures when the rheological behavior of the fluids is described by the Williamson–Cross model:

$$k_l a^* = F_{wc,r}\left(U_g^* , Fr , \mu^* , \sigma^* , n_w , t_w^* \right) \tag{6.75}$$

where $F_{N,r}$, $F_{ost,r}$ and $F_{wc,r}$ are the process relationships required.

The reduced configuration of the system and the operating points explored are represented, for each case, in Figure 6.39.

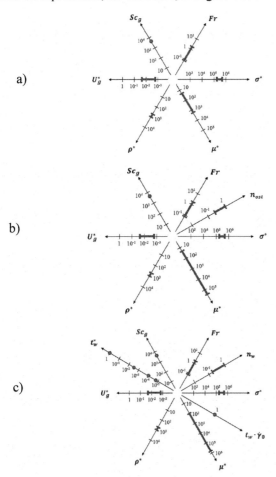

Figure 6.39. *Study of gas–liquid mass transfer in a stirred tank: graphical representation of the reduced configuration of the system: a) only Newtonian fluids, b) Newtonian and shear-thinning fluids described by the Ostwald-de-Waele model, c) Newtonian and shear-thinning fluids described by the Williamson–Cross model*

First, the process relationship $F_{N,r}$ is determined for *Newtonian fluids*. Hassan *et al.* [HAS 12] show that $F_{N,r}$ can be approximated by the following monomial function:

$$k_l a^* = 0.21 \times \left(Fr \cdot U_g^* \right)^{2/3} \cdot \left(\mu^* \right)^{-0.59} \cdot \left(\sigma^* \right)^{-0.25} \qquad [6.76]$$

A graphical representation, where the impact of σ^* is hidden, can be used to display the effect of the viscosity of Newtonian fluids (μ^*) on the target internal measure $k_l a^*$, whatever the other variables are. Represented in Figure 6.40, it shows that, for a given rotation speed and gas flow rate (U_g^* and Fr), an increase in μ^* leads to a clear reduction in the performance of the gas–liquid mass transfer.

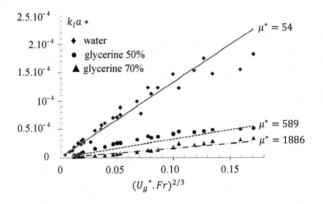

Figure 6.40. *Study of gas–liquid mass transfer in a stirred tank: effect of the internal measure μ^* on $k_l a^*$ in the case of Newtonian fluids. Each symbol corresponds to a different Newtonian fluid*

Let us now examine process relationships for *shear-thinning fluids*.

For this, we will first consider the case where the rheological behavior of fluids is described by the material function associated with the *Ostwald-de-Waele model*. Of course, this model is valid for Newtonian fluids: in this case, $k = \mu_l$ and $n_{ost} = 1$.

The process relationship $F_{ost,r}$ required (equation [6.74]) is approximated by a monomial function so that the exponents on the internal measures, other than n_{ost}, are identical to those obtained for the single Newtonian fluids (equation [6.76]). This leads to an exponent equal to -1.34 on n_{ost}. The experimental target internal measures and the ones predicted by this new process relationship are compared in Figure 6.41. A poor agreement is obtained whatever the shear-thinning fluids. This clearly shows that the addition of a single internal measure, n_{ost}, to the initial relevance list of physical quantities influencing $k_l a^*$ does not allow us to group all the points into a single master curve, in this case, the curve corresponding to Newtonian fluids. The material function associated with the Ostwald-de-Waele model is therefore not adapted to describe the influence of the variability of the viscosity of the shear-thinning fluids tested.

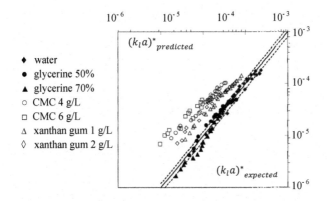

Figure 6.41. *Study of the gas–liquid mass transfer in a stirred tank where shear-thinning fluids are described by the Ostwald-de-Waele model: comparison between the experimental target internal measures and those predicted by the process relationship (equation [6.74])*

Let us now examine the case where the material function associated with the *Williamson–Cross model* is considered. This model is applied to Newtonian fluids: in this case, $t_w^* = 1$, $n_w = 1$ and $\mu_w = 2\mu_l$. Hassan *et al.* [HAS 12] show that the process relationship $F_{wc,r}$ (equation [6.75]) can be approximated by the following monomial function:

$$k_l a^* = 0.021 \times \left(Fr \cdot U_g^* \right)^{2/3} . \left(\mu^* \right)^{-0.59} . \left(\sigma^* \right)^{-0.25} . \left(n_w \right)^{-2.40} . \left(t_w^* \right)^{-0.17} \qquad [6.77]$$

The target internal measures $k_l a^*$ both experimental and predicted in equation [6.77] are compared in Figure 6.42(a). A good level of adequacy is obtained this time, whether Newtonian or shear-thinning fluids are considered.

A graphical representation (Figure 6.42(b)) can also be proposed to display the effect of the internal measure t_w^* on $k_l a^*$, regardless of the values of the other internal measures. It can be observed that the points associated with each shear-thinning fluid are remarkably grouped along the lines corresponding to the values of t_w^* which decrease when they move further away from the points corresponding to the Newtonian fluids, for which $t_w^* = 1$.

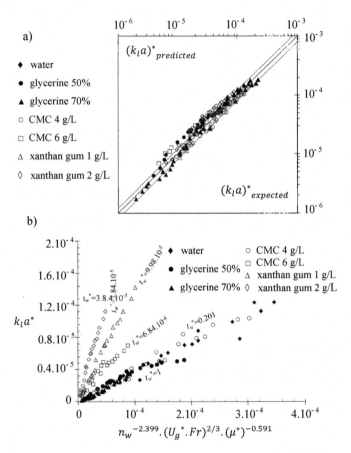

Figure 6.42. *Study of the gas–liquid mass transfer in a stirred tank where pseudoplastic fluids are described by the Williamson–Cross model: a) comparison between the experimental target internal measures and those predicted by the process relationship (equation [6.77]), b) effect of the internal measure t_w on $k_l a^*$*

6.5.4. *Conclusion*

This example shows:

– that an over-simplification of the relevance list of parameters characterizing the behavior of a shear-thinning fluid leads to bias in the construction of the set of internal measures which influence $k_l a^*$;

– that a single process relationship able to predict the dimensionless volumetric gas–liquid mass transfer coefficient $k_l a^*$ in the presence of Newtonian and shear-thinning fluids can be obtained to correctly define the configuration of the material (introduction of two internal measures n_w and t_w^*);

– that it is possible, using a rigorous theoretical framework, to overcome the Metzner–Otto concept (1957) [MET 57] in order to define the reference shear rate.

The process relationship obtained was established:

– for the range of dimensionless numbers shown in Figure 6.39(c);

– for the geometry of the tank, impeller and distributor in this study.

In this form, this process relationship does not help in ascertaining how the geometric parameters which influence $k_l a^*$. Other tests will be necessary, especially in order to evaluate the effect of the modified Schmidt number: indeed, equation [6.77] was only obtained for a value of this number.

6.6. Ohmic heating

Continuous heat treatment of complex food products is mainly carried out using conventional heat exchangers. These involve heated walls and thermosensitive fluids that can be altered when they come into contact with them (organoleptic modifications, undesirable reactions such as the denaturation of proteins), which often leads to clogging problems and a heterogeneous heat treatment of the fluid.

In this context, heating using the direct Joule effect, also called ohmic heating, is an interesting alternative to conventional processes, due to its energy efficiency, homogeneity in temperature generated (especially for heterogeneous products) and the compact nature of the equipment [QUA 95]. Furthermore, it helps to eliminate hot surfaces, thus significantly reducing fouling problems. It consists of passing an electric current directly into the product flowing between a pair of electrodes. Despite its advantages, industrial use of ohmic heating remains limited, mainly because the criteria necessary for scaling-up have not be established.

The example which we will present in this section aims to show that dimensional analysis is able to accurately establish the conditions of similarity which need to be respected in order to extrapolate an ohmic heating process. To do so, the dependence of the fluid's electric conductivity on temperature must be examined.

6.6.1. Description of the problem

This example focuses on a process involving a jet ohmic heater. This innovative technology helps to overcome hydraulic problems (residence time dispersion) commonly encountered in tubular ohmic heat exchangers [GHN 08]. In this device, an electrical current is passing in the jet of fluid flowing continuously between two electrodes. The jet acts as an electrical resistance and the product is heated by the direct Joule effect.

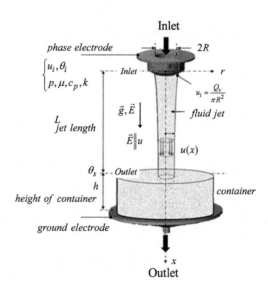

Figure 6.43. *Diagram of a jet ohmic heating process*

As shown in Figure 6.43, the problem can be described in the following way. A circular injection nozzle with a radius R ejects a jet of Newtonian fluid with a density ρ and a viscosity μ at a mass flow rate Q. Using an initial mean velocity u_i, this jet falls freely into a container, and it is thus accelerated by the gravitational field. The distance between the injection nozzle and the container is denoted by L. The length of the jet and the electrical power dissipated varies according to the height h of the container. An alternating voltage V is applied between the two electrodes, generating an electric field $\vec{E}$ parallel to the field of velocity $\vec{u}$ of the product, thus heating the fluid jet with an electrical conductivity k by the Joule effect.

The goal of this study is to identify the dimensionless numbers governing the outlet temperature of the jet of fluid (θ_{out}) according to the operating conditions (length of the jet, flow rate of the fluid, electrical current and electrical conductivity).

To this end, the following simplified hypotheses will be considered:

– the steady electrical force induced is negligible since an alternating electrical field is applied to an incompressible Newtonian fluid with a density of charge of zero;

– the viscous dissipation in the product is negligible with respect to heat generation by the Joule effect. The current I measured at inlet gives a faithful representation of the electrical power delivered [GHN 08a, GHN 09b];

– the velocity of the fluid in the jet is independent from the radius (piston flow), and thus only depends on the axial coordinate: $u(x)$;

– the initial or inlet temperature of the product, θ_i, in the jet cell is uniform;

– due to their low dependence on temperature, the density ρ and heat capacity c_p of the fluid are assumed to be constant;

– the effect of viscosity on heat transfer may be neglected since there is no friction from walls (free jet) and the Reynolds number is large at the outlet of the jet.

6.6.2. *Target variable and list of relevant independent physical quantities*

6.6.2.1. If the electrical conductivity of the fluid is constant

The target variable is the outlet temperature of the fluid jet just before it falls into the container (Figure 6.43): it is denoted by θ_{out}.

The physical quantities influencing θ_{out} can be broken down into:

– *boundary conditions*: these are the following geometric parameters: the radius of the injection nozzle R and the length of the jet L[8];

– *material parameters*: these are the density of the Newtonian fluid ρ, its specific heat capacity c_p and its electrical conductivity k;

– *process parameters*: these are the inlet temperature of the fluid θ_i, the electrical current I and the fluid velocity at the inlet of the nozzle u_i, which is an intermediate variable defined by:

$$u_i = \frac{Q_v}{\pi.R^2} \qquad [6.78]$$

where Q_v is the volumetric flow rate of the fluid;

– *universal constants*: this is the gravitational acceleration g.

Therefore, when the electrical conductivity of the fluid is considered as being constant, the problem can be given in terms of physical quantities influencing the outlet temperature of the fluid jet (θ_{out}) whereby:

$$\theta_{out} = f_1(R, L, \rho, c_p, k, \theta_i, I, u_i, g) \qquad [6.79]$$

---

8 Strictly speaking, a set of geometric parameters $\{p_{geo}\}$, other than R and L, which completely describe the geometry of the jet ohmic heating process, should also be listed. We decided from the start not to list them because a single geometry of the process will be tested in the experimental program.

6.6.2.2. If the dependence of the fluid's electrical conductivity on temperature is taken into account

The variable physical property of the material is the fluid's electrical conductivity k which depends on the temperature θ. The dimensional material function considered in this example is therefore: $s(p) = k(\theta)$.

Traditionally, $k(\theta)$ can be modeled by a linear relationship with the following form (we will return to this point during the description of the experimental program):

$$k(\theta) = k_1.(1 + c.\theta) \qquad [6.80]$$

where k_1 is a constant dimensionally homogeneous to an electrical conductivity (S.m^{-1}) and c the temperature factor (°C^{-1})[9]. These two parameters depend on the nature of the fluid.

According to the rules given in Chapter 4, the initial relevance list of physical quantities influencing the outlet temperature of the fluid jet, θ_{out}, established considering constant electrical conductivity (equation [6.79]) must be supplemented by:

– the reference temperature, denoted by θ_0;

– the electrical conductivity at the reference temperature, $k_0 = k(\theta_0)$;

– the set of dimensionless numbers, denoted as $\{\pi_m\} = \{a_0.\theta_0\}$, which appears in the argument u of the standard dimensionless material function w except ratio (θ / θ_0).

The dimensional material function $k(\theta)$ as described in equation [6.80] is reference-invariant in so far as it takes the following form:

$$k(\theta) = (A + B.\theta)^C \qquad [6.81]$$

with $A = k_1$, $B = k_1.c$ and $C = 1$.

---

9 We decided, for practical reasons, to use degrees Celsius. We could have used kelvins without altering any subsequent developments.

As a result, the reference temperature θ_0 must not be added to the initial relevance list.

Nevertheless, it is necessary to choose a value of θ_0 in order to calculate $k_0 = k(\theta_0)$. For reasons of simplicity, let us consider θ_0 as equal to the fluid's inlet temperature θ_i (of course, any other choice would also have been possible):

$$\theta_0 = \theta_i \tag{6.82}$$

According to the discussion in Chapter 4, the standard dimensionless material function associated with $k(\theta)$, $w(u)$, can be expressed here as:

$$w(u) = \frac{k(\theta)}{k(\theta_0)} \tag{6.83}$$

And its argument u is defined as:

$$u = a_0.(\theta - \theta_0) = \frac{1}{k(\theta_0)}.\left(\frac{dk(\theta)}{d\theta}\right)_{\theta=\theta_0}.(\theta - \theta_0) \tag{6.84}$$

Let us now calculate u. According to equation [6.80], the derivative of the fluid's electrical conductivity in terms of the temperature at $\theta = \theta_0$ is:

$$\left(\frac{dk(\theta)}{d\theta}\right)_{\theta=\theta_0} = c.k_1 \tag{6.85}$$

hence:

$$a_0 = \frac{c.k_1}{k_1.(1+c.\theta_0)} = \frac{c}{1+c.\theta_0} \tag{6.86}$$

This therefore gives:

$$a_0.\theta_0 = \frac{c.\theta_0}{1+c.\theta_0} \tag{6.87}$$

It is interesting to note that $a_0.\theta_0$ can be expressed according to the dimensionless number $c.\theta_0$. As a result, the set of dimensionless numbers $\{\pi_m\} = \{a_0.\theta_0\}$ to add to the initial relevance list in this example is:

$$\{\pi_m\} = \{c.\theta_0\} \tag{6.88}$$

Therefore, in the case where the dependence of electrical conductivity on temperature is taken into account, the problem can be written as:

$$\theta_{out} = f_2(R, L, \rho, c_p, k_0 = k(\theta_0), \theta_i, I, u_i, g, c.\theta_0) \tag{6.89}$$

Equation [6.89] therefore includes an additional physical quantity compared to the case where the electrical conductivity of the fluid is considered as being constant (equation [6.79]).

6.6.3. *Establishing dimensionless numbers and the configuration of the system*

The associated dimensional matrix, when electrical conductivity depends on temperature, is shown in Figure 6.44. It can be noted that:

– five fundamental dimensions are necessary to express the dimensions of the physical quantities listed: temperature (K), length (L), mass (M), time (T) and ampere (A);

– according to the Vaschy–Buckingham theorem, six internal measures (11 physical quantities – 5 fundamental dimensions) will describe the process relationship;

– the repeated variables chosen are θ_i, R, ρ, u_i and $k_0 = k(\theta_0)$. However, any other choice could have been taken as long as the rules given in Chapter 2 were respected.

	θ_{out}	I	g	L	c_p	$c.\theta_0$	θ_i	R	ρ	u_i	k_0
K	1	0	0	0	-1	0	1	0	0	0	0
L	0	0	1	1	2	0	0	1	-3	1	-3
M	0	0	0	0	0	0	0	0	1	0	-1
T	0	0	-2	0	-2	0	0	0	0	-1	-1
A	0	1	0	0	0	0	0	0	0	0	2

Figure 6.44. *Jet ohmic heating process: dimensional matrix*

The modified dimensional matrix, obtained following the transformation of the core matrix into an identity matrix is shown in Figure 6.45.

θ_{out}	I	g	L	c_p	$c.\theta_0$	θ_i	R	ρ	u_i	k_0
1	0	0	0	-1	0	1	0	0	0	0
0	3/2	-1	1	0	0	0	1	0	0	0
0	1/2	0	0	0	0	0	0	1	0	0
0	3/2	2	0	2	0	0	0	0	1	0
0	1/2	0	0	0	0	0	0	0	0	1

Figure 6.45. *Jet ohmic heating process: modified dimensional matrix*

The set of internal measures is obtained using the coefficients contained in the modified residual matrix, and is expressed as follows:

$$\left\{ \begin{array}{l} \pi_1 = \dfrac{\theta_{out}}{\theta_i} \;;\; \pi_2 = \dfrac{I}{R^{3/2}.\rho^{1/2}.u_i^{3/2}.\sigma_0^{1/2}} \;;\; \pi_3 = \dfrac{g}{R^{-1}.u_i^2} \\[3mm] \pi_4 = \dfrac{L}{R} \;;\; \pi_5 = \dfrac{c_p}{\theta_i^{-1}.u_i^2} \;;\; \pi_6 = c.\theta_0 \end{array} \right\} \qquad [6.90]$$

After rearrangements of the dimensionless numbers, this set becomes:

$$\left\{ \begin{array}{l} \pi_1 = \dfrac{\theta_{out}}{\theta_i} \;;\; \pi_2 = \dfrac{I^2}{R^3.\rho.u_i^3.\sigma_0} \;;\; \pi_3 = \dfrac{u_i^2}{g.R} \\[3mm] \pi_4 = \dfrac{L}{R} \;;\; \pi_5 = \dfrac{c_p.\theta_i}{u_i^2} \;;\; \pi_6 = c.\theta_0 \end{array} \right\} \qquad [6.91]$$

Therefore, dimensional analysis transforms equation [6.89] into:

$$\frac{\theta_{out}}{\theta_i} = F\left(\pi_2 = \frac{I^2}{\rho.u_i^3.k_0.R^3} \;;\; \pi_3 = \frac{u_i^2}{g.R} \;;\; \pi_4 = \frac{L}{R} \;;\; \pi_5 = \frac{c_p.\theta_i}{u_i^2} \;;\; \pi_6 = c.\theta_i \right) \quad [6.92]$$

where F is the process relationship describing the evolution of θ_{out} in a jet ohmic heating process, and $\pi_6 = c.\theta_i$ since we chose a reference temperature of $\theta_0 = \theta_i$ (equation [6.82]). The second term of equation [6.92] represents the configuration of the system.

6.6.4. *Analytical expression of the process relationship*

Previously, we carried out a "blind" dimensional analysis of the jet ohmic heating process, in which the variation of the fluid's electrical conductivity with temperature was taken into account. This allowed us to rigorously define the configuration of the system (equation [6.92]).

Let us now compare this configuration to the existing analytical solution to predict the evolution of the fluid jet's outlet temperature θ_{out} which is the process relationship F. The full demonstration leading to this analytical solution was carried out by Delaplace *et al.* [DEL 09]. If we consider an element of the volume of the jet with a length of dx and a cross-sectional area of dA, the enthalpic balance in this elementary volume can be formulated as follows:

$$Q.c_p.\mathrm{d}\theta = \sigma_1.(1+k.\theta).E^2.A(x).\mathrm{d}x = \frac{I^2}{\sigma_1.(1+k.\theta).A(x)} \cdot \mathrm{d}x \qquad [6.93]$$

where Q is the mass flow of the fluid (kg.s^{-1}), E is the electric field generated (V.m^{-1}), x the abscissa along the fluid jet (m), θ the temperature within the fluid jet at a given position x and $A(x)$ is the longitudinal section of the jet assumed circular. By integrating this enthalpic balance along the length of the jet and taking account of the boundary conditions:

– $\theta = \theta_i$ at the outlet of the nozzle ($x = 0$);

– $\theta = \theta_{out}$ at the impact of the jet into the container ($x = L$);

we obtain:

$$\theta_{out} = \frac{\left[\left[1+2k.\left(\frac{\frac{k.\theta_i^2}{2}+\theta_i+\left(\frac{I^2.u_i}{3\rho.\pi^2.R^4.\sigma_1.c_p}\right)}{\times\left(\left(1+\frac{2g}{u_i^2}.\frac{L}{R}\right)^{3/2}-1\right)}\right)\right]^{1/2}-1\right]}{k} \qquad [6.94]$$

This equation can be rewritten to include the dimensionless numbers $\left(\pi_1,\pi_2,\pi_3,\pi_4,\pi_5,\pi_6\right)$ defined in equation [6.92], hence:

$$\frac{\theta_{out}}{\theta_i} = \frac{1}{\pi_6} \left[\left[1 + \pi_6^2 + 2\pi_6 \cdot \left(\frac{1 + \left(\frac{\pi_2.(1+\pi_6).\pi_3}{3\pi^2.\pi_5} \right)}{\times \left(\left(1 + \frac{2}{\pi_3}.\pi_4 \right)^{3/2} - 1 \right)} \right) \right]^{1/2} - 1 \right] \qquad [6.95]$$

Therefore, the analytical solution expressed in equation [6.95], which is nothing else than the process relationship required F, perfectly agrees with the configuration established by "blind" dimensional analysis (equation [6.92]): the target internal measure (θ_{out}/θ_i) depends on five other internal measures, π_2 to π_6. Equation [6.95] also shows that:

– the contribution of the dimensionless number $\pi_6 = k.\theta_i$ (material configuration), confirming that the variation of the electrical conductivity of the fluid with temperature can in no way be neglected;

– the process relationship characteristic of this jet ohmic heating process does not have a monomial form[10].

6.6.5. Validation of the process relationship

The experimental device and program used by Ghmini *et al.* [GHN 09] helped to validate the process relationship (equation [6.92]). In this study of the jet ohmic heating process:

– the diameter of the injection nozzle ($2R$) was a space is needed betwen 0.013 m with a jet length of L varying from 0.07 to 0.17 m;

– the experimental parameters measured were the inlet and outlet temperatures in each zone, the current, the supply voltage and the total power of the electrical supply;

– the profile of temperatures along the fluid jet were measured by an infrared camera and a contact thermometer;

---

10 We should point out that at this stage there is nothing surprising about this. Indeed, Chapter 2 showed that the monomial form was the mathematically form generally considered due to its ability to describe families of various functions, and that nothing guarantees that the process relationship can be put in a monomial form or can adjust to the "true" physical law.

– the model fluids used were aqueous solutions of NaCl at three different concentrations, whose electrical conductivity were measured and the parameters c and k_l included in equation [6.80] were identified;

– the mass flows of the fluids varied between 170 and 300 kg.h^{-1}, corresponding to inlet speeds of the fluid u_i between 0.35 and 0.63 m.s^{-1};

– the heating power varied between 1.8 and 4 kW (corresponding to electrical intensities varying between 1.1 and 1.8 A);

– the height of the container varied between 0 and 30 mm.

The configuration of the system and the operating points explored are represented in Figure 6.46.

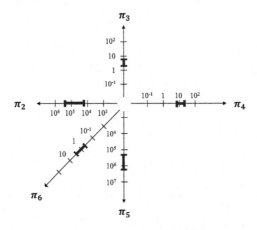

Figure 6.46. *Jet ohmic heating process: configuration of the system and operating points explored*

This study showed that the analytical model helps to faithfully describe the experimental evolution of the temperature along the jet, since the gap between the experimental points systematically remains below 5%. This is confirmed by Figure 6.47, showing the overall temperature increases (denoted by $\Delta\theta = \theta_{out} - \theta_i$) predicted by the process relationship (equation [6.94]) as a function of the experimental overall temperature increases. A good level of adequacy is achieved. The predicted temperature increases are slightly higher than those measured, which can be explained partly by the heat losses of the ohmic heating system and partly by the accuracy of the measurement instruments (temperatures and electrical quantities).

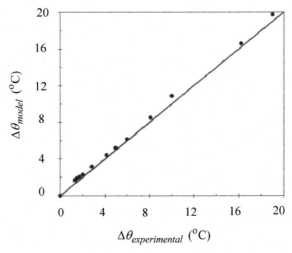

Figure 6.47. *Jet ohmic heating: overall temperature increases*
($\Delta\theta = \theta_{out} - \theta_i$) predicted by the process relationship as a function of the
experimental overall temperature increases

6.6.6. Conclusion

This example shows that the dimensional analysis of an ohmic heating process must take account the dependence of electrical conductivity on temperature in order to define the complete set of internal measures responsible for the temperature of the fluid at the outlet. The configuration of the system established using dimensional analysis is in line with the analytical solution proposed by Delaplace *et al.* [DEL 09] to predict the axial profile of the temperature along the fluid jet. The experiments carried out by these authors helped to validate the process relationship obtained in this way, which do not have a monomial form.

These findings provide the theoretical framework which should be applied for scaling the electrical power to dissipate in order to achieve the desired heating of the fluid according to the operating conditions in place.

Conclusions

In this book, we have presented the theoretical framework and the principles of the dimensional analysis process. We showed how they can be used for taking into account the variability of a material's physical properties within the transformation operation. We have also highlighted the guidelines to follow and the tools available for rigorously building a process relationship (semi-empirical correlation) between dimensionless numbers.

Dimensional analysis, associated with the theory of similarity and experiments on laboratory-scale equipment (models), remains a powerful tool although it is currently underused in food process engineering. We hope that this book will contribute to its rehabilitation, especially through the various examples taken from our research which show the range of possibilities offered by dimensional analysis to:

– establish the causality relationship between the final characteristics of the product and the operating conditions governing the process;

– identify the optimal operating points for a transformation operation;

– transpose the results from one scale to another.

Figure C.1 shows a flowchart of the entire dimensional analysis process, and for each step provides the chapters to which it refers.

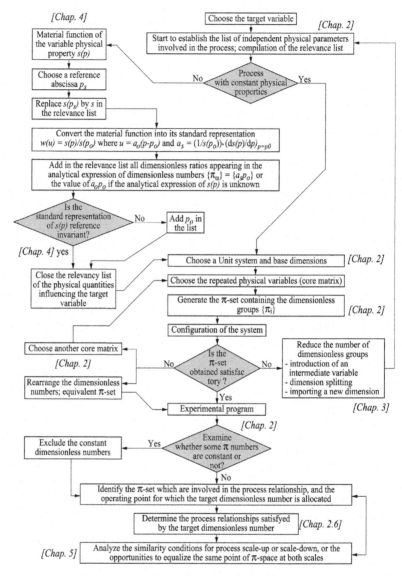

Figure C.1. *Flowchart of the dimensional analysis process*

Moreover, as stated a number of times throughout the book, each of these stages is governed by theoretical considerations which, however rigorous they may be, leave the users free to make their own choices (e.g. introducing intermediate variables, choosing the base, rearranging dimensional numbers, etc.). Given these choices, an uninitiated user may encounter some difficulties (e.g. establishing the relevance list of physical quantities). On this issue, Cussler [CUS 97] quite rightly states:

> I want to emphasize that most people go through three mental states concerning this method. At first they believe it is a route to all knowledge, a simple technique by which any set of experimental data can be greatly simplified. Next they become disillusioned when they have difficulties in the use of the technique. These difficulties commonly result from efforts to be too complete. Finally, they learn to use the method with skill and caution, benefiting both from their past successes and from their frequent failures. I mention three stages because I am afraid many may give up at the second stage and miss the real benefits involved.

In reality, the only option is to persevere. Indeed, the user of laboratory-scale equipment must absolutely know how to standardize the problem with dimensionless numbers in order to mimic the operating point of the targeted industrial installation, identify a physical model of phenomena and transpose the results obtained onto another scale. Dimensional analysis and the theory of similarity unquestionably provide a single theoretical framework for carrying out the extrapolation and transposition of the results, without any recourse to risky methods and possible pitfalls of simple intuition. Indeed, unfortunately, "less knowledge means more convictions" [CYR 02].

Furthermore, another original contribution of this book (and undoubtedly its prime motivation) is to show how to take into account the variability of a material's physical properties in a dimensional analysis process. We hope that the theoretical elements defined in Chapter 4 and the examples in Chapter 6 will have convinced the user of the relevance and interest of this technique, and that they will thus become more commonly applied.

Dimensional analysis and numerical simulations are often placed in opposition to each other. The latter approach, which is more deterministic, can provide a result for a given configuration of the system and information on the influence of parameters. However, this is often at the cost of a relatively significant calculation time. They provide results on a local scale (e.g. a velocity or concentration field) which are difficult to use on a more global scale, in particular for the scaling of the process.

A completely different approach is also traditionally proposed by scientists who are not inclined toward modeling and simulation. This consists of implementing an experimental design. Although this is a way of reducing the amount of experiments, this approach only produces a set of polynomials, without a particular physical meaning which helps us to find the optimal operating point. Moreover, this optimum is only valid for the equipment tested, which limits the contribution and generic value of the results obtained.

Is it possible to combine and make these three approaches work together?

In certain configurations where numerical simulation works, could we not define the set of numerical experiments to carry out by using an experimental design? These results of numerical experiments would then be processed by dimensional analysis to expand their scope by making them generic and easier to use (process relationship). Indeed, numerical experiments are still easier to carry out than the real experiments. Since they are nevertheless longer and more expensive to carry out than may be expected, it makes sense to reduce them by choosing the correct conditions to test them in. Making them more generic and bringing in knowledge garnered from dimensional analysis paints a perfect picture ... which now needs to be explored.

Appendices

Appendix 1

Shift Relationships between Spaces of Dimensionless Numbers

Chapter 2 (section 2.1.3.4.3) showed that there exists shift relationships between two spaces of dimensionless numbers obtained for two sets of repeated variables chosen. The objective of this appendix is to illustrate this by referring back to example 2.5 used extensively in Chapter 2.

The base chosen in example 2.5 (denoted as "base 1") corresponds to the repeated variables V_3, V_4 and V_5. The dimensional matrix associated with this base was given in Figure 2.2, the modified dimensional matrix (denoted by $\mathbf{D}_{m, \text{base } 1}$) in Figure 2.3 and the space of dimensionless numbers associated (denoted by $\pi_{i, \text{base } 1}$) by equation [2.22]. Now, we will refer to "base 2", which is another choice of repeated variables, for example V_4, V_2 and V_5. This will produce the space of dimensionless numbers denoted by $\{\pi_{i, \text{base } 2}\}$. The dimensional matrix associated with base 2 (V_4, V_2 and V_5) is given in Figure A1.1.

	V_1	V_3	V_6	V_4	V_2	V_5
d_1	0	1	2	1	0	1
d_2	-2	-3	1	0	-1	-1
d_3	1	2	0	-3	0	-1

Figure A1.1. *Example 2.5: dimensional matrix (black outline) associated with base 2. The core matrix associated with the repeated variables (V_4, V_2 and V_5) is given in the dark gray boxes and the associated residual matrix is given in light gray boxes*

The modified dimensional matrix associated with base 2 will be named $D_{m,\ base\ 2}$. The first column in Figure A1.2 shows the linear combinations necessary to obtain this matrix $D_{m,\ base\ 2}$.

	V_1	V_3	V_6	V_4	V_2	V_5
$-0.5d_1-0.5d_3$	-0.5	-1.5	-1	1	0	0
$-1.5d_1-d_2-0.5d_3$	1.5	0.5	-4	0	1	0
$1.5d_1+0.5d_3$	0.5	2.5	3	0	0	1

Figure A1.2. *Example 2.5: modified dimensional matrix $D_{m,\ base\ 2}$ (black outline) associated with base 2. The associated core identity matrix $C_{I,\ base\ 2}$ is in dark gray and the modified residual matrix $R_{m,\ base\ 2}$ is in light gray*

The space of dimensionless numbers, denoted by $\{\pi_{i,base2}\}$ associated with this modified dimensional matrix $D_{m,\ base\ 2}$, is given by:

$$\{\pi_{i,base2}\} = \begin{cases} \pi_{1,base2} = \dfrac{V_1}{V_4^{-0.5}.V_2^{1.5}.V_5^{0.5}} , \pi_{2,base2} \\[2ex] = \dfrac{V_3}{V_4^{-1.5}.V_2^{0.5}.V_5^{2.5}} , \pi_{3,base2} = \dfrac{V_6}{V_4^{-1}.V_2^{-4}.V_5^{3}} \end{cases} \qquad [A1.1]$$

We will now show that it is possible to shift from the set of dimensionless numbers $\{\pi_{i,base\ 1}\}$ to the set of dimensionless numbers $\{\pi_{i,base\ 2}\}$. To do so, matrix **A** and sub-matrices S_1 and S_3 must be constructed using the rules given in Chapter 2 (section 2.1.3.4.3). They are shown in Figure A1.3.

Succession of columns in D with base 2 (V_4, V_2, V_5)

		V_1	V_3	V_6	V_4	V_2	V_5
	V_1	1	0	0	0	0	0
	V_2	0	0	0	0	1	0
	V_6	0	0	1	0	0	0
	V_3	0	1	0	0	0	0
	V_4	0	0	0	1	0	0
	V_5	0	0	0	0	0	1

Succession of columns in D with base 1 (V_3, V_4, V_5)

Figure A1.3. *Example 2.5: matrix **A** constructed using two bases. Matrices S_1 and S_3 are in light gray and dark gray, respectively*

Using the knowledge of the sub-matrices S_1 and S_3 (Figure A1.3), and of the modified residual matrix associated with base 1 $R_{m, \text{ base 1}}$ (Figure 2.3), the shift matrix **P** can be calculated using equation [2.24]:

$$\mathbf{P} = \left[S_1 - R_{m, \text{base 1}}^{\text{T}} \times S_3 \right]^{-1}$$

$$\mathbf{P} = \left[\begin{bmatrix} 1 & 0 & 0 \\ 0 & 0 & 0 \\ 0 & 0 & 1 \end{bmatrix} - \begin{bmatrix} 3 & 2 & -8 \\ 4 & 3 & -13 \\ -7 & -5 & 23 \end{bmatrix}^{T} \times \begin{bmatrix} 0 & 1 & 0 \\ 0 & 0 & 0 \\ 0 & 0 & 0 \end{bmatrix} \right]^{-1}$$

$$\mathbf{P} = \left[\begin{bmatrix} 1 & 0 & 0 \\ 0 & 0 & 0 \\ 0 & 0 & 1 \end{bmatrix} - \begin{bmatrix} 3 & 4 & -7 \\ 2 & 3 & -5 \\ -8 & -13 & 23 \end{bmatrix} \times \begin{bmatrix} 0 & 1 & 0 \\ 0 & 0 & 0 \\ 0 & 0 & 0 \end{bmatrix} \right]^{-1} \qquad [A1.2]$$

$$\mathbf{P} = \left[\begin{bmatrix} 1 & 0 & 0 \\ 0 & 0 & 0 \\ 0 & 0 & 1 \end{bmatrix} - \begin{bmatrix} 0 & 3 & 0 \\ 0 & 2 & 0 \\ 0 & -8 & 0 \end{bmatrix} \right]^{-1} = \begin{bmatrix} 1 & -3 & 0 \\ 0 & -2 & 0 \\ 0 & 8 & 1 \end{bmatrix}^{-1} = \begin{bmatrix} 1 & -1.5 & 0 \\ 0 & -0.5 & 0 \\ 0 & 4 & 1 \end{bmatrix}$$

Equations [2.23] and [A1.2] help to calculate the set of dimensionless numbers $\{\pi_{i,base\,2}\}$, whereby:

$$\mathbf{P} \times \begin{bmatrix} \ln\left(\pi_{1,base\,1}\right) \\ \ln\left(\pi_{2,base\,1}\right) \\ \ln\left(\pi_{3,base\,1}\right) \end{bmatrix} = \begin{bmatrix} \ln\pi_{1,base\,1} - 1.5\ln\pi_{2,base\,1} \\ -0.5\ln\pi_{2,base\,1} \\ 4\ln\pi_{2,base\,1} + \ln\pi_{3,base\,1} \end{bmatrix}$$

$$= \begin{bmatrix} \ln\dfrac{\pi_{1,base\,1}}{\left(\pi_{2,base\,1}\right)^{1.5}} \\ \ln\dfrac{1}{\left(\pi_{2,base\,1}\right)^{0.5}} \\ \ln\left[\left(\pi_{2,base\,1}\right)^{4}\left(\pi_{3,base\,1}\right)\right] \end{bmatrix} = \begin{bmatrix} \ln\dfrac{V_1}{V_4^{-0.5}.V_2^{1.5}.V_5^{0.5}} \\ \ln\dfrac{V_3}{V_4^{-1.5}.V_2^{0.5}.V_5^{2.5}} \\ \ln\dfrac{V_6}{V_4^{-1}.V_2^{-4}.V_5^{3}} \end{bmatrix} = \begin{bmatrix} \ln\left(\pi_{1,base\,2}\right) \\ \ln\left(\pi_{2,base\,2}\right) \\ \ln\left(\pi_{3,base\,2}\right) \end{bmatrix} \qquad [\text{A1.3}]$$

Eventually, the result obtained, which is the set of dimensionless numbers $\{\pi_{i,base2}\}$, agrees with the one obtained in equation [A1.1].

Appendix 2

Physical Meaning of Dimensionless Numbers Commonly Used in Process Engineering

The objective of this appendix is to provide the physical meaning of the dimensionless numbers commonly used in process engineering, which are often the ones looked for when a blind dimensional analysis is carried out. To do this, the basic principles of configurational analysis are used. The theoretical developments which help, using conservation equations, to express different fluxes of mass, momentum or energy, and therefore the associated internal measures, will not be given in this appendix. For more details, we invite the reader to consult [BEC 76] or [MID 81].

In the configurational analysis, the system being studied is the one defined in section 2.3 of Chapter 2 (Figure 2.4). The physical quantity Ψ taken into account is of three different types:

– total or partial mass of species A (dimension: [M]; unit SI: kg);

– momentum (dimension: $[M.L.T^{-1}]$; unit SI: $kg.m.s^{-1}$);

– energy (dimension: $[M.L^2.T^{-2}]$; unit SI: joule).

It is possible to express the fluxes associated with each of these physical quantities, using the integral form of the conservation equations, general physical laws (e.g. Fourier's law) and/or laws specific to interfaces (e.g. Newton's law for heat transfer). From there, an internal measure of each of these fluxes can be defined, which is a ratio between the two fluxes, ϕ_i and ϕ_j:

$$\pi_{ij} = \frac{\phi_i}{\phi_j} \qquad\qquad [A2.1]$$

A2.1. Internal measures related to the material fluxes

The main internal measures associated with the material fluxes are summarized in Table A2.1, which we have adapted from [BEC 76] and [MID 81]. In this table, the names and conventions chosen are as follows:

– Δw_A and ΔC_A are, respectively, the characteristic variation of the mass fraction of chemical species A [-] and the concentration of chemical species A [M.L^{-3}] so that $\rho.\Delta w_A \equiv \Delta C_A$ [1];

– t can be the time taken since the start of a process [T], a time constant of the process or any other characteristic time;

– r_A is a characteristic reaction rate of chemical species A [M.L^{-3}.T^{-1}];

– U is the mean flow velocity in the system [M. T^{-1}];

– δ is the thickness of a boundary layer [L] located along the boundary surface S which is active in the process;

– k is a transfer coefficient (or conductance) [L.T^{-1}];

– D_A and D are the molecular diffusion coefficient of species A and a generalized diffusion coefficient [L^2.T^{-1}];

– the expressions of material flux involving different geometrical variables (volume V, surface S or the thickness layer δ) are, depending on the case, replaced by V/S or substituted by:

$$V \cong L^3, S \cong L^2, \delta \cong L \qquad\qquad [A2.2]$$

where L is the characteristic length of the system.

– the traditional names of the different ratios obtained and their physical meaning are specified.

---

1 The notation "$\cong$" means "measured by", thus showing the idea that the term preceding this symbol can be approximated by the simplified expression given after this symbol.

Material flux ϕ_i / Material flux ϕ_j	Accumulation	Convection	Dispersion	Reaction
Material flux ϕ_j	$\dfrac{\rho.V.\Delta w_A}{t}$ or $\dfrac{\rho.\delta.S.\Delta w_A}{t}$	$\rho.U.S.\Delta w_A$ or $k.\rho.S.\Delta w_A$	$\dfrac{D.\rho.S.\Delta w_A}{L}$ or Diffusion $\dfrac{D_A.\rho.S.\Delta w_A}{\delta}$	$r_A \cdot V$ or $r_A \cdot S \cdot \delta$
Accumulation $\dfrac{\rho.V.\Delta w_A}{t}$ or $\dfrac{\rho.\delta.S.\Delta w_A}{t}$	1	$\dfrac{U.t}{L}$ *Thomson*[1]	$\dfrac{D_A.t}{L^2}$ *Fourier*[2]	$\dfrac{r_A.t}{\Delta C_A}$ *Damköhler*[3]
Convection $\rho.U.S.\Delta w_A$ or $k.\rho.S.\Delta w_A$	–	$\dfrac{k}{U}$ *Stanton*[4]	–	$\dfrac{r_A.L}{U.\Delta C_A}$ *Damköhler I*[5] $\dfrac{r_A.D_A}{k^2.\Delta C_A}$ *Hatta*[6]
Dispersion / Diffusion $\dfrac{D.\rho.S.\Delta w_A}{L}$ or $\dfrac{D_A.\rho.S.\Delta w_A}{\delta}$	–	$\dfrac{U.L}{D_A}$ *Péclet*[7] $\dfrac{U.L}{D}$ *Bodenstein*[8] $\dfrac{k.L}{D_A}$ *Sherwood*[9] $\dfrac{k.L}{D_e}$ *Biot*[10]	1	$\dfrac{r_A.L^2}{D_A.\Delta C_A}$ *Damköhler II*[11]
Reaction $r_A \cdot V$ or $r_A \cdot S \cdot \delta$	–	–	–	1

Table A2.1. *Internal measures related to material fluxes*

1 The *Thomson number (Th)* is used for unsteady flows (e.g. turbulence structures generated behind an obstacle). It compares the convective transport with the material accumulation, and can therefore be seen as the ratio between the characteristic time of unsteady flows and the convection time (L/U). *Th* is the opposite of the *Strouhal number (Str)* in which the characteristic time of unsteady flows is replaced by a frequency: $Str = f.L/U$.

2 The *Mass Fourier number* (Fo_M) compares the purely diffusional transport and the material accumulation. It is encountered in the study of mass transfer in a transient regime.

3 The *Damköhler number (Da)* examines a characteristic chemical reaction time and the time *t* passed since the inlet of the fluid into a reactor.

4 The *material Stanton number* (St_M) compares the effective mass transport over a fluid boundary layer and the mass transport by convection (residence time).

5 The *Damköhler I number (Da_I)* is applied here to a continuous type plug flow reactor. It compares the flux of species A consumed by the chemical reaction and the mass flux transported by the flow (L/U is the same as the residence time).

6 When the convective transport is replaced by the transport over a layer, the most widely known form of the Damköhler I number is the *Hatta number (Ha)* in which the transfer conductance, *k*, is expressed using film theory: $k = D_A / \delta$. It is widely used in fluid-fluid reactors.

7 The *material Péclet number* (Pe_M) compares the convective transport with the transport by molecular diffusion (D_A). Depending on the characteristic dimension considered, it will be axial or transversal. It is interesting to note that $Pe_M=Sc.Re$ where *Sc* and *Re* are the Schmidt number (see section A2.4) and the Reynolds number (see Table A2.2), respectively. Finally, it can also be observed that in chemical engineering, the material Péclet number is commonly defined not by using the molecular diffusion coefficient D_A but a dispersion coefficient *D*: in this case, it is the same as the Bodenstein number.

8 The *Bodenstein number (Bo)* compares the transport by convection with the transport by dispersion (since the coefficient *D* includes molecular diffusion, the dispersion by gradients of velocity and/or turbulent dispersion). For instance, it is used to take account of the phenomena of dispersion inside the chemical reactors.

9 The *Sherwood number* (*Sh*) compares the transport of mass over a fluid boundary layer with purely diffusional mass transport (D_A).

10 The *material Biot number* (Bi_M) has a physical meaning which is close to the Sherwood number, but is applied as a preference to the mass transport from both sides of the surface of a solid body (particle of powder during drying, catalyzer, resin exchangers, ions, etc.). It involves an effective mass diffusion coefficient (D_e), and thus examines the material fluxes outside and inside a solid body.

11 The *Damköhler II number* (Da_{II}) is used in the calculation of tubular chemical reactors. It compares the term related to the chemical reaction with dispersion or diffusion transport. In the case of heterogeneous catalytic reactors, it takes the form of the *Thiele number* and compares the flux of species A consumed by the chemical reaction and the flux transferred by diffusion inside the catalyzer (it also involves an effective mass diffusion coefficient D_e).

A2.2. Internal measures related to the momentum fluxes

The various terms in the conservation equation of momentum can be seen as the *momentum fluxes* or *forces*.

The main internal measures associated with these fluxes are summarized in Table A2.2, which we have adapted from [BEC 76] and [MID 81]. It is constructed in the same way as Table A2.1, and also involves the following parameters:

$-p$ is the dynamic pressure [$M.L^{-1}.T^{-2}$];

$-g$ is the gravitational acceleration [$L.T^{-2}$];

$-\tau_w$ is the shear stress at the wall [$M.L^{-1}.T^{-2}$];

$-F_D$ is the total drag force [$M.L.T^{-2}$];

$-\gamma$ is the interfacial tension [$M.T^{-2}$] and R is the radius of the curved interface in the expression of the capillary force.

In Table A2.2, only the most commonly used dimensionless numbers have been named, and their physical meaning is explained.

Momentum flux ϕ_i Momentum flux ϕ_j	Inertial force $\dfrac{\rho.U.L^3}{t}$	Convection $\rho.U^2.L^2$	Buoyancy force $g.L^3.\Delta\rho$	Driving pressure $L^2.\Delta p$	Viscous force $\mu.U.L$	Friction force at wall $L^2.\tau_w$	Total drag force F_D	Capillary force $\gamma.L$
Inertial force $\dfrac{\rho.U.L^3}{t}$	1	$\dfrac{U.t}{L}$ Thomson[12]	$\dfrac{g.t.\Delta\rho}{\rho.U}$	$\dfrac{t.\Delta p}{\rho.U.L}$	—	$\dfrac{t.\tau_w}{\rho.U.L}$	$\dfrac{F_D.t}{\rho.U.L}$	$\dfrac{\gamma.t}{\rho.U.L^2}$
Convection $\rho.U^2.L^2$	—	1	—	$\dfrac{\Delta p}{\rho.U^2}$ Euler, Cavitation[13]	—	$\dfrac{\tau_w}{\rho.U^2}$ Friction coefficient[14]	$\dfrac{F_D}{\rho.U^2.L^2}$ Drag coefficient[14]	—
Buoyancy force $g.L^3.\Delta\rho$	—	$\dfrac{\rho.U^2}{g.L.\Delta\rho}$ Froude[15]	1	$\dfrac{\Delta p}{g.L.\Delta\rho}$	$\dfrac{\mu.U}{g.L^2.\Delta\rho}$	$\dfrac{\tau_w}{g.L.\Delta\rho}$	$\dfrac{F_D}{g.L^3.\Delta\rho}$	—
Driving pressure $L^2.\Delta p$	—	—	—	1	$\dfrac{\mu.U}{L.\Delta p}$	$\dfrac{\tau_w}{\Delta p}$	$\dfrac{F_D}{L^2.\Delta p}$	$\dfrac{\gamma}{L.\Delta p}$

Viscous force $\mu.U.L$	$\dfrac{\rho.L^2}{\mu.t}$ Stokes[16]	$\dfrac{\rho.U.L}{\mu}$ Reynolds[17]	—	—	1	$\dfrac{L.\tau_w}{\mu.U}$	$\dfrac{F_D}{\mu.U.L}$	—
Friction force at wall $L^2.\tau_w$	—	—	—	—	—	1	$\dfrac{F_D}{L^2.\tau_p}$	$\dfrac{\gamma}{L.\tau_p}$
Total drag force F_D	—	—	—	—	—	—	1	—
Capillary force $\gamma.L$	—	$\dfrac{\rho.U^2.L}{\gamma}$ Weber[18]	$\dfrac{g.L^2.\Delta\rho}{\gamma}$ Bond[19]	—	$\dfrac{\mu.U}{\gamma}$ Capillary[20]	—	$\dfrac{F_D}{\gamma.L}$	1

Table A2.2. *Internal measures related to momentum fluxes (or forces)*

12 The *Thomson number* (*Th*) is encountered in transient flow problems (see section A2.1).

13 The *Euler number* (*Eu*) is a measure of pressure losses due to friction (e.g. in a pipe). If this is caused by the occurrence of a singularity (bend, fitting, valve, etc.), it corresponds to the *singular coefficient* (ξ). When the pressure loss is an overpressure exceeding the vapor tension, it refers to the *cavitation number* (σ_C).

14 The *coefficients of friction* (*C_f*) and *drag* (*C_D*) are of the same nature. They translate the resistance caused by the interaction force between a fluid and a solid surface into the convective momentum transport. C_f is used for flows in ducts and C_D is used for wake phenomena around obstacles or inclusions (bubble, droplet, particle, etc.).

15 The *Froude number* (*Fr*) compares the convective momentum transport to the buoyancy force, $g.L^3.\Delta\rho$. Nevertheless, it is generally defined with respect to the weight, $g.L^3.\rho$, and is expressed as $Fr=U^2/g.L$. It is involved in free surface or multi-phase flows, when gravity may have a separation effect on the components of the different densities.

16 The *Stokes number* (*Sto*) is used in fluid dynamics to study the behavior of a particle in a fluid (e.g. sedimentation). It compares the inertia of a solid particle (accumulation of momentum) with the viscous forces. In general, ρ is the density of the particle, L is its diameter and t is a relaxation time. If $Sto \ll 1$, the solid particle may be considered as a tracer of the fluid flow.

17 The *Reynolds number* (*Re*) compares the momentum flux by convection (inertial force) with the flux of momentum by viscous diffusion (viscous force). It helps to identify the flow regime (laminar, transitory or turbulent) in the system being studied.

18 The *Weber number* (*We*) compares the convective transport of momentum (inertial force) with the interfacial tension force (capillary force). It is very widely used in the study of dispersed systems (e.g. formation or deformation of fluid particle).

19 The *Bond number* (*Bo*) compares the buoyancy force (gravitational effects) with the capillary force in the case of fluid interfaces.

20 The *capillary number* (*Ca*) compares the viscous force with the interfacial tension force (e.g. two-phase flow in a microchannel).

A2.3. Internal measures related to the energy fluxes

The main internal measures in thermal energy fluxes[2] are summarized in Table A2.3, adapted from [BEC 76] and [MID 81]. It is constructed in the same way as Tables A2.1 and A2.2, and also involves the following parameters:

– c_p is the specific heat capacity $[L^2.T^{-2}.K^{-1}]$.

– λ is the thermal conductivity $[M.L.T^{-3}.K^{-1}]$.

– ε_s is the emissivity of a solid body [-], σ is the Stefan–Boltzmann constant ($5.67.10^{-8}$ W.m^{-2}.K^{-4}) and $<T>$ is the mean temperature [K].

– h_C is the heat exchange coefficient (conductance) by convection $[M.T^{-3}.K^{-1}]$.

– $(\Delta H_r.r_A)$ is the rate of thermal energy generation by the chemical reaction of species A per unit of volume $[M \cdot L^{-1} \cdot T^{-3}]$ since ΔH_r is the reaction enthalpy per unit of mass $[L^2.T^{-2}]$.

---

2 The energy conservation equation in open systems comes from the application of the first law of thermodynamics. It involves the enthalpy transport, the accumulation of internal energy and the various terms of heat generation (chemical reactions and work of mechanical friction forces). The term "kinetic energy transport" is redundant with respect to the momentum balance, in which it is already included: it is thus eliminated from this balance. In the end, only the internal energy conservation equation is considered in its integral form.

Energy flux ϕ_i / Energy flux ϕ_j	Accumulation	Convection	Conduction, dispersion	Radiation	Chemical reaction	Friction	Compression	Interfacial transfer
Energy flux ϕ_i	$\dfrac{\rho c_p L^3 \Delta T}{t}$	$\rho c_p U L^2 \Delta T$	$\lambda . L . \Delta T$	$\sigma \varepsilon_s L^2 T^3 \Delta T$	$\Delta H_r r_A L^3$	$\mu . L . (\Delta U)^2$	$\dfrac{L^3 . \Delta P}{t}$	$h_c . L^2 . \Delta T$
Accumulation $\dfrac{\rho c_p L^3 \Delta T}{t}$	1	$\dfrac{U t}{L}$ Thomson	$\dfrac{\alpha . t}{L^2}$ Fourier[21]	$\dfrac{\sigma . \varepsilon_s . T^3 . t}{\rho . c_p . L}$	$\dfrac{\Delta H_r . r_A . t}{\rho . c_p . \Delta T}$	$\dfrac{\mu . (\Delta U)^2 . t}{\rho . c_p . \Delta T . L^2}$	$\dfrac{\Delta P}{\rho . c_p . \Delta T}$	$\dfrac{h_c . t}{\rho . c_p . L}$
Convection $\rho c_p U L^2 \Delta T$	—	1	—	—	$\dfrac{\Delta H_r . r_A . L}{\rho . c_p . U . \Delta T}$ Damköhler III[22]	$\dfrac{\mu . U}{\rho . c_p . L . \Delta T}$	$\dfrac{\Delta P . L}{\rho c_p U . \Delta T . t}$	$\dfrac{h_c}{\rho . c_p . U}$ Stanton[23]
Conduction, dispersion $\lambda . L . \Delta T$	—	$\dfrac{U . L}{\alpha}$ Péclet[24]	1	$\dfrac{\sigma . \varepsilon_s . L . T^3}{\lambda}$	$\dfrac{\Delta H_r . r_A . L^2}{\lambda . \Delta T}$ Damköhler IV[25]	$\dfrac{\mu . U^2}{\lambda . T}$ Brinkman[26]	$\dfrac{\Delta P . L^2}{\lambda . \Delta T . t}$	$\dfrac{h_c . L}{\lambda}$ Nusselt[27] $\dfrac{h_c . L}{\lambda_e}$ Biot[28]

Radiation	$\mathscr{E}_s L^2 T^3 \Delta T$	—	$\dfrac{\rho . c_p . U}{\sigma . \mathscr{E}_s . T^3}$ Thring[29]	—	1	$\dfrac{\Delta H_r . r_A . L^2}{\sigma . \mathscr{E}_S . T^3 . \Delta T}$	$\dfrac{\mu . U^2}{\mathscr{E}_s L T^3 \Delta T}$	$\dfrac{\Delta P . L}{\mathscr{E}_s T^3 \Delta T . t}$	$\dfrac{h_c}{\sigma . \mathscr{E}_s . T^3}$
Chemical reaction	$\Delta H_r . r_A L^3$	—	—	—	—	1	$\dfrac{\mu . U^2}{\Delta H_r . r_A . L^2}$	$\dfrac{\Delta P}{\Delta H_r . r_A . t}$	$\dfrac{h_C . \Delta T}{\Delta H_r . r_A . L}$
Friction	$\mu . L . (\Delta U)^2$	—	—	—	—	—	1	$\dfrac{L^2 . \Delta P}{\mu . U^2 . t}$	$\dfrac{L . h_C . \Delta T}{\mu . U^2}$
Compression	$\dfrac{L^3 . \Delta P}{t}$	—	—	—	—	—	—	1	$\dfrac{h_C . \Delta T . t}{L . \Delta P}$
Interfacial transfer	$h_C . L^2 . \Delta T$	—	—	—	—	—	—	—	1

Table A2.3. *Internal measures related to the energy fluxes*

21 The *thermal Fourier number* (Fo_T) is the same as the mass Fourier number, but the molecular diffusivity $\mathcal{D}_A$ is replaced by thermal diffusivity

$$\alpha = \frac{\lambda}{\rho.c_p}.$$

22 The *Damköhler III number* (Da_{III}) represents the ratio between the thermal flux generated by the chemical reaction and the thermal energy transport by convection.

23 The *thermal Stanton number* (St_T), just as with mass transfer Stanton number, compares the transport of thermal energy over a fluid boundary layer and the transport of thermal energy by convection.

24 In the same way as with mass transfer, the *thermal Péclet number* (Pe_T) compares the transport of thermal energy by convection with the transport of thermal energy by diffusion. Length L is the characteristic dimension of the reactor or the particle. We should note that there is no equivalent to the Bodenstein number in heat transfer and that $Pe_T=Pr.Re$ where Pr and Re are the Prandtl number (see section A2.4) and the Reynolds number (see Table A2.2), respectively.

25 The *Damköhler IV number* (Da_{IV}), also called the *Pomerantsev number*, is equivalent to the Damköhler II number. It represents the ratio between the flux of thermal energy generated by the chemical reaction and the transport of thermal energy by diffusion or dispersion.

26 The comparison of the viscous generation of thermal energy (friction) with the flux of thermal energy transported by diffusion is described by the *Brinkman number* (Br). It is often used for laminar flows with high shear velocities.

27 The *Nusselt number* (Nu) is equivalent to the Sherwood number. It compares the transport of thermal energy over a fluid boundary layer (e.g. in a pipe wall) with purely diffusional transport (conduction in a fluid).

28 The *thermal Biot number* (Bi_T), equivalent to the material Biot number, has a physical meaning which is close to the Nusselt number, but is applied in preference to heat transfer on either side of the surface of a solid body (e.g. particle of powder during drying). It involves an effective thermal conductivity (λ_e) and therefore, represents the competition between the thermal energy transfer fluxes inside and outside the solid body.

29 The *Thring number* (*Th*) compares the convective transport of thermal energy with the thermal energy flux caused by radiation. It is the opposite of the *Bansen number*. It can also be expressed in terms of the thermal coefficient by radiation h_R, which is equivalent to the product $\sigma \varepsilon_s T^3$.

A2.4. Other internal measures

We saw in Chapter 2 (section 2.3) that the configuration of a system is defined in terms of flux, as well as in terms of retention and geometric domain. Internal measures are also associated with these retentions and geometric domains, some of which are examined below.

A2.4.1. *Internal measures of retention*

The retentions of a system relate to its physical quantities which we used to express the various fluxes: the total mass (or the mass of species A), momentum and energy. The internal measures of retention are rarely named and used as they are in the studies involving dimensional analysis, but they are present. For instance, we could cite the *volume fraction of the dispersed phase* in a two-phase system or the *mass or molar fraction* of species A in a mixture. Midoux [MID 81] also mentions certain dimensionless numbers connected to thermodynamics:

– the *Arrhenius number* (ratio between the activation energy of a process and the thermal agitation energy) defined by $Ar = \dfrac{E}{R.T}$;

– the *Carnot yield* (ratio between the thermal energy which can be transformed into mechanical energy and total thermal energy) defined by $\dfrac{T_h - T_c}{T_h}$ (T_h and T_c being hot and cold source temperatures, respectively).

There are other internal measures of retention, but it is impossible to provide an exhaustive list of them.

A2.4.2. *Geometric internal measures*

The geometric description of the configuration of a system is given using three types of quantities: length, surface and volume. The geometric internal measures are, therefore, generally presented in the form of ratios between two characteristic lengths, surfaces or volumes.

Among the ratio of the two characteristic lengths, we can give D/L the ratio between the diameter D of a pipe (or a long object) and its length L, H/D the ratio between the height H of a column and its diameter D, d/T the ratio between the diameter of an impeller d and the diameter of the stirred tank T.

A well-known relationship of surfaces is the *sphericity* of a particle, which is defined as the ratio between the surface of a sphere of the same volume and the surface of the particle.

There are many other geometric internal measures which could also be defined.

It should be noted that it is often difficult to establish a complete list of geometric parameters able to perfectly describe the geometry of a system (see Chapter 6). In practice, only a few geometric internal measures are taken into account: those whose role is considered to be significant on the process. This explains why the process relationships obtained are generally affected by the geometry in which they have been established.

A2.4.3. *Composite internal measures*

We refer here to the terminology used by Midoux [MID 81] to name dimensionless numbers displaying several phase properties. For instance, numbers which compare the fluxes of quantities of different natures (mass, momentum and heat), such as:

– the *Schmidt number*:

$$Sc = \frac{convective \ \ mass \ \ flux}{diffusive \ \ mass \ \ flux} \times \frac{diffusive \ \ momentum \ \ flux}{convective \ \ momentum \ \ flux}$$

$$Sc = \frac{Pe_M}{Re} = \frac{\rho.U.L^2.\Delta w_A}{D_A.\rho.L^2.\Delta w_A / L} \times \frac{\mu.U.L}{\rho.L^2.U^2} = \frac{\mu}{D_A.\rho}$$

[A2.3]

– the *Prandtl number*:

$$Pr = \frac{convective \quad heat \quad flux}{diffusive \quad heat \quad flux} \times \frac{diffusive \quad momentum \quad flux}{convective \quad momentum \quad flux}$$

$$Pr = \frac{Pe_T}{Re} = \frac{\rho.c_p.U.L^2.\Delta T}{\lambda.L.\Delta T} \times \frac{\mu.U.L}{\rho.L^2.U^2} = \frac{c_p.\mu}{\lambda}$$

[A2.4]

– the *Lewis number*:

$$Le = \frac{convective \quad mass \quad flux}{diffusive \quad mass \quad flux} \times \frac{diffusive \quad heat \quad flux}{convective \quad heat \quad flux}$$

$$Le = \frac{Pe_M}{Pe_T} = \frac{\rho.U.L^2.\Delta w_A}{D_A.\rho.L^2.\Delta w_A / L} \times \frac{\lambda.L.\Delta T}{\rho.c_p.U.L^2.\Delta T} = \frac{\lambda}{\rho.c_p.D_A}$$

[A2.5]

Appendix 3

The Transitivity Property of the Standard Non-dimensionalization Method and its Consequences on the Mathematical Expression of Reference-Invariant Standard Dimensionless Material Functions (RSDMFs)

As highlighted in Chapter 4, certain standard dimensionless material functions, known as reference-invariant, do not depend on the reference abscissa. The aim of this appendix is to show that this is directly linked to the transitivity property of the standard non-dimensionalization method. We will demonstrate that, by using this property, it is possible to define the mathematical form of the standard reference-invariant dimensionless material functions. This group of material functions has interesting consequences, especially that of reducing the number of variables to introduce into the relevance list of independent physical quantities influencing the target variable.

A3.1. Transitivity of the standard non-dimensionalization method

A3.1.1. *Definitions*

Let us consider the transformation of a material function $s(p)$ according to the standard non-dimensionalization method (see Chapter 4, section 4.1), but with two different reference abscissas, denoted by p_1 and p_2. In Figure A3.1, the function $s(p)$ is represented, as well as the two chosen reference points, $(p_1;s(p_1))$ and $(p_2;s(p_2))$. These two reference points enable us to obtain two standard dimensionless material functions of the function $s(p)$, represented by $w_1(u_1)$ and $w_2(u_2)$, respectively (Figure A3.2). u_1, w_1, u_2 and w_2 are defined as follows:

$$\begin{cases} u_1(p) = \dfrac{(p-p_1)}{s(p_1)}\cdot\left(\dfrac{ds}{dp}\right)_{p=p_1} & \text{and} \quad w_1 = \dfrac{s(p)}{s(p_1)} \\[4mm] u_2(p) = \dfrac{(p-p_2)}{s(p_2)}\cdot\left(\dfrac{ds}{dp}\right)_{p=p_2} & \text{and} \quad w_2 = \dfrac{s(p)}{s(p_2)} \end{cases} \qquad [\text{A3.1}]$$

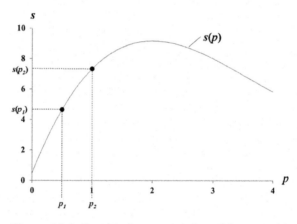

Figure A3.1. *Graphical representation of the material function s(p) (from [PAW 91])*

By definition, we have $u_1(p_1) = 0$ and $w_1(p_1) = 1$. Subsequently, the point with coordinates $(0;1)$ of the standard dimensionless material function w_1 is the image of point $(p_1;s(p_1))$.

In the same way, we have $u_2(p_2) = 0$ and $w_2(p_2) = 1$: the point with coordinates $(0;1)$ of the standard dimensionless material function w_2 is the image of point $(p_2;s(p_2))$.

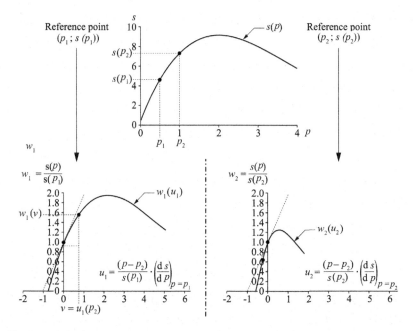

Figure A3.2. *Graphical representation of the standard dimensionless material functions w₁ and w₂ of the material function s(p)*

If we also consider $w_1(u_1)$ as a material function, it is again possible to apply the standard non-dimensionalization method to it. By choosing any reference abscissa, denoted by v, the new standard dimensionless material function, w_3 of the argument u_3, is written as:

$$u_3(v) = \frac{(u_1 - v)}{w_1(v)} \cdot \left(\frac{dw_1}{du_1}\right)_{u_1 = v} \quad \text{and} \quad w_3 = \frac{w_1(u_1)}{w_1(v)} \tag{A3.2}$$

If we define the reference abscissa v so that $v = u_1(p_2)$, then this gives:

$$v = \frac{(p_2 - p_1)}{s(p_1)} \cdot \left(\frac{ds}{dp}\right)_{p = p_1} \quad \text{and} \quad w_1(v) = \frac{s(p_2)}{s(p_1)} \tag{A3.3}$$

Therefore, the image of point $\left(p_2\,;s(p_2)\right)$ by the standard dimensionless material function w_1 is the point $\left(v\,;w_1\left(v\right)\right)$ (Figure A3.3).

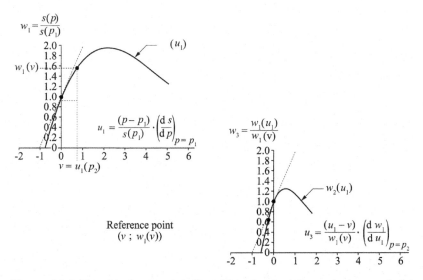

Figure A3.3. *Graphical representation of the standard dimensionless material function w_1 and the standard dimensionless material function w_3*

When we have:

– a dimensional material function $s(p)$;

– from which are taken two standard dimensionless material functions, w_1 and w_2, respectively, defined at the reference abscissas p_1 and p_2;

– and a standard dimensionless material function   obtained by applying the standard non-dimensionalization method to function w_1, at the reference abscissa $v=u_1\left(p_2\right)$.

This transformation of the function s into function w is called *transitive* when function w_2 is identical to function w_3 (Figure A3.4).

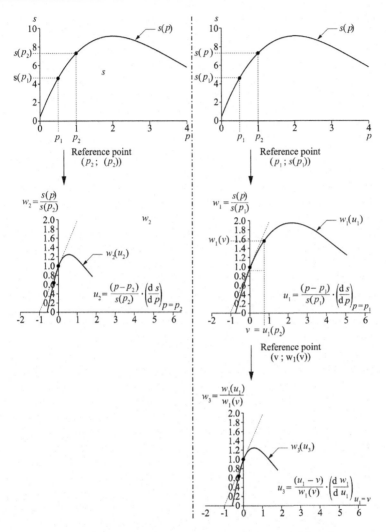

Figure A3.4. *Property of transitivity of the material function*

A3.1.2. *Transitivity of the standard dimensionless method*

It is therefore necessary to show that "w_2" and "w_1 *followed by* w_3" are identical, whereby for each point $(p;s(p))$, they coincide: $u_3 = u_2$ and $w_3 = w_2$.

By replacing u, v and w_1 with their expressions (equations [A3.1] and [A3.3]) in the expression of u_3 (equation [A3.2]), we get:

$$
u_3 = \frac{(u_1 - v)}{w_1(v)} \cdot \left(\frac{dw_1}{du_1}\right)_{u_1 = v} = \left(\frac{(p - p_1)}{s(p_1)} \cdot \left(\frac{ds}{dp}\right)_{p = p_1} - \frac{(p_2 - p_1)}{s(p_1)} \cdot \left(\frac{ds}{dp}\right)_{p = p_1}\right) \cdot
$$
$$
\frac{s(p_1)}{s(p_2)} \cdot \left(\frac{dw_1}{du_1}\right)_{u_1 = v}
$$
[A3.4]

Hence:

$$
u_3 = \left(\frac{(p - p_2)}{s(p_1)} \cdot \left(\frac{ds}{dp}\right)_{p = p_1}\right) \cdot \frac{s(p_1)}{s(p_2)} \cdot \left(\frac{dw_1}{du_1}\right)_{u_1 = v}
$$
[A3.5]

We can break down $\left(\dfrac{dw_1}{du_1}\right)_{u_1 = v}$ into:

$$
\left(\frac{dw_1}{du_1}\right)_{u_1 = v} = \left(\frac{\dfrac{dw_1}{dp}}{\dfrac{du_1}{dp}}\right)_{u_1 = v}
$$
[A3.6]

The expression $\dfrac{dw_1}{dp}$ is obtained using that of w_1 (equation [A3.1]):

$$
\left(\frac{dw_1}{dp}\right)_{u_1 = v} = \frac{1}{s(p_1)} \cdot \left(\frac{ds}{dp}\right)_{p = p_2}
$$
[A3.7]

since $v = u_1(p_2)$.

By combining equations [A3.6] and [A3.7], we get:

$$
\left(\frac{dw_1}{du_1}\right)_{u_1 = v} = \frac{1}{s(p_1)} \cdot \left(\frac{ds}{dp}\right)_{p = p_2} \cdot \frac{1}{\left(\dfrac{du_1}{dp}\right)_{p = p_2}}
$$
[A3.8]

Moreover, the expression of u_1 (equation [A3.1]) shows that its derivative to p is a constant:

$$\frac{du_1}{dp} = \frac{d}{dp}\left(\frac{p}{s(p_1)}\cdot\left(\frac{ds}{dp}\right)_{p=p_1} - \frac{p_1}{s(p_1)}\left(\frac{ds}{dp}\right)_{p=p_1}\right) = \frac{1}{s(p_1)}\cdot\left(\frac{ds}{dp}\right)_{p=p_1} \qquad [A3.9]$$

By combining equations [A3.5], [A3.8] and [A3.9], we get:

$$u_3 = \left(\frac{(p-p_2)}{s(p_1)}\cdot\left(\frac{ds}{dp}\right)_{p=p_1}\right)\cdot\frac{s(p_1)}{s(p_2)}\cdot\frac{1}{s(p_1)}\left(\frac{ds}{dp}\right)_{p=p_2}\cdot\frac{1}{\frac{1}{s(p_1)}\cdot\left(\frac{ds}{dp}\right)_{p=p_1}} \qquad [A3.10]$$

which leads to:

$$u_3 = \frac{(p-p_2)}{s(p_2)}\cdot\left(\frac{ds}{dp}\right)_{p=p_2} \qquad [A3.11]$$

The previous expression corresponds to the definition of u_2 (equation [A3.1]); therefore, $u_3 = u_2$. Furthermore, we have:

$$w_3 = \frac{w_1(u_1)}{w_1(v)} = \frac{\dfrac{s(p)}{s(p_1)}}{\dfrac{s(p_2)}{s(p_1)}} = \frac{s(p)}{s(p_2)} = w_2 \qquad [A3.12]$$

This proves the transitivity of this transformation.

A3.1.3. *Comment*

The functions obtained by the standard method are not independent of the chosen reference abscissas. Indeed, in the previous developments, w_1 and w_2 do not coincide along the entire range of u, but only at $u = 0$, likewise for w_1 and w_3 (Figure A3.4). Obtaining functions that are independent of the reference abscissas requires an additional condition to be applied. This is the objective of the following section.

A3.2. Applying the property of transitivity

A3.2.1. Definition of reference-invariant standard dimensionless function

Let us consider two standard dimensionless functions $w_1(u)$ and $w_2(u)$ of the same material function obtained by taking any two different reference points $(p_1; s(p_1))$ and $(p_2; s(p_2))$, respectively (Figure A3.5). Let us imagine that these representations $w_1(u)$ and $w_2(u)$ are juxtaposed so that, whatever the value of u, $w_1(u) = w_2(u)$.

Furthermore, the property of transitivity implies that there is a standard representation $w_3(u)$ of $w_1(u)$ which is itself juxtaposed to $w_2(u)$: $w_2(u) = w_3(u)$. We can deduce from this that $w_1(u)$ and $w_3(u)$ are identical. In other words, this means that if two standard transformations of the same material function are identical, any standard transformation of one or the other will be juxtaposed to this representation, without any special consideration being given to the chosen point of reference.

Conversely, let us now consider two standard dimensionless functions so that (Figure A3.5):

– $w_1(u)$ is the standard representation of the function s, obtained by taking $(p_1; s(p_1))$ as a reference point;

– $w_3(u)$ is the standard representation of the function $w_1(u)$ obtained by taking any reference point, named here $(v; w_1(v))$.

Let us assume that these consecutive standard functions can be juxtaposed so that, whatever the value of u, $w_1(u) = w_3(u)$. The property of transitivity means that the standard function $w_3(u)$ is itself juxtaposed to a standard dimensionless function of the function s, $w_2(u)$ obtained by taking $(p_2; s(p_2))$ as a point of reference which is the antecedent of $(v = u_1(p_2); w_1(v))$ by the function $w_1 : w_2(u) = w_3(u)$.

We can deduce from this that $w_1(u)$ and $w_2(u)$ are themselves identical, even though they are obtained at any two different points of reference $(p_1; s(p_1))$ and $(p_2; s(p_2))$.

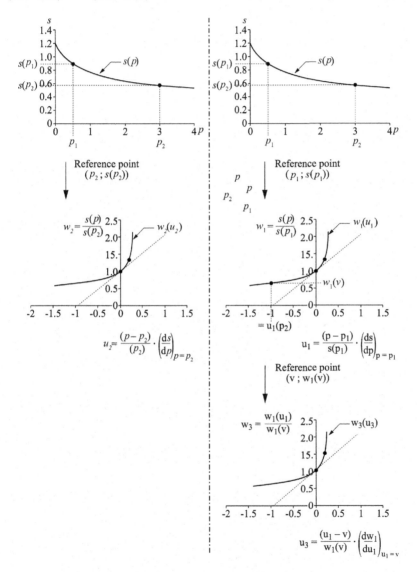

Figure A3.5. *Standard transformations of a reference-invariant material function*

Consequently, if a standard dimensionless function w_1 from a material function s and its own standard transformation ("w_1 followed by w_3") are juxtaposed, then any standard dimensionless function from s with any reference abscissas will produce a unique dimensionless material function, independent of the reference abscissa chosen. This function is called the *reference-invariant standard dimensionless material function (denoted by ϕ) to distinguish it from a non-reference-invariant standard dimensionless material function (denoted by w)*.

A3.2.2. Mathematical forms of reference-invariant standard dimensionless material functions

Once these elements are defined, we can examine the mathematical form of these reference-invariant standard dimensionless functions. This means determining which types of material functions will satisfy the following condition: "*the standard dimensionless material function $w_1 = \phi_1(u_1)$ from s and the standard dimensionless material function $w_3 = \phi_3(u_3)$ from w_1, defined at any reference points, are identical*". Such an identity does not mean that there is a point-to-point invariance in the representation, but that the points $\left(u_1; w_1(u_1)\right)$ and their representations $\left(u_3; w_3(u_3)\right)$ must be juxtaposed to the same invariant curve. For instance, it is clear from Figure A3.5 that the representations of w_1 and w_3 align along the same curve without a strict correspondence with the images of points $\left(p_1; s(p_1)\right)$ and $\left(p_2; s(p_2)\right)$.

According to equation [A3.2], $\varphi_3(u_3)$ is written as:

$$\phi_3(u_3) = \frac{\phi_1(u_1)}{\phi_1(v)}$$ [A3.13]

with v as the reference abscissa at which ϕ_3 is defined. The property of transitivity means that it is possible to write:

$$\phi_2 = \phi_3$$ [A3.14]

The condition of invariance demands:

$$\phi_2 = \phi_1 \tag{A3.15}$$

If the invariance of a function is established, it must necessarily conform to equations [A3.14] and [A3.15], which then gives:

$$\phi_1 = \phi_2 = \phi_3 = \phi \tag{A3.16}$$

ϕ subsequently represents the reference-invariant standard dimensionless material function resulting from the equality of [A3.16].

Equation [A3.13] can thus be rewritten as:

$$\phi(u_3).\phi(v) - \phi(u_1) = 0 \tag{A3.17}$$

with:

$$u_3 = \frac{(u_1 - v)}{\phi(v)} \cdot \left(\frac{d\phi}{du_1}\right)_{u_1 = v} \tag{A3.18}$$

Let us define (u_i) $(i \in (1,3))$ as the normalized variation of the standard material function ϕ:

$$u_i \mapsto g(u_i) = \frac{1}{\phi(u_i)} \cdot \left(\frac{d\phi}{du_i}\right) \tag{A3.19}$$

Therefore, this gives:

$$g(u_1) = \frac{1}{\phi(u_1)} \cdot \left(\frac{d\phi}{du_1}\right) \tag{A3.20}$$

We can thus rewrite expression [A3.18] as:

$$u_3 = (u_1 - v).g(v) \tag{A3.21}$$

Let us derive [A3.17] in relation to v since $\phi(u_1)$ does not depend on v, hence:

$$\frac{\partial u_3}{\partial v} \cdot \frac{\partial \phi(u_3)}{\partial v} \cdot \phi(v) + \phi(u_3) \cdot \frac{\partial \phi(v)}{\partial v} = 0 \qquad [A3.22]$$

We can also obtain the expression of $\dfrac{\partial u_3}{\partial v}$ by deriving [A3.21] in relation to v:

$$\frac{\partial u_3}{\partial v} = -g(v) + (u_1 - v) \cdot \frac{\partial g(v)}{\partial v} \qquad [A3.23]$$

Hence, by replacing $(u_1 - v)$ with its expression [A3.21]:

$$\frac{\partial u_3}{\partial v} = -g(v) + \frac{u_3}{g(v)} \cdot \frac{\partial g(v)}{\partial v} \qquad [A3.24]$$

The combination of equations [A3.22] and [A3.24] leads to:

$$\left(-g(v) + \frac{u_3}{g(v)} \cdot \frac{\partial g(v)}{\partial v} \right) \cdot \frac{\partial \phi(u_3)}{\partial v} \cdot \phi(v) + \phi(u_3) \cdot \frac{\partial \phi(v)}{\partial v} = 0 \qquad [A3.25]$$

By using equation [A3.19], we can write:

$$\frac{\partial \phi(v)}{\partial v} = g(v) \cdot \phi(v) \qquad [A3.26]$$

which can thus be replaced in [A3.25] :

$$\left(-g(v) + \frac{u_3}{g(v)} \cdot \frac{\partial g(v)}{\partial v} \right) \cdot \frac{\partial \phi(u_3)}{\partial v} \cdot \phi(v) + \phi(u_3) \cdot \phi(v) \cdot g(v) = 0 \qquad [A3.27]$$

By dividing by $\phi(u_3) \cdot \phi(v)$ expression [A3.27], we obtain:

$$\left(-g(v) + \frac{u_3}{g(v)} \cdot \frac{\partial g(v)}{\partial v} \right) \cdot \frac{\partial \phi(u_3)}{\partial v} \cdot \frac{1}{\phi(u_3)} + g(v) = 0 \qquad [A3.28]$$

Hence, by combining the previous equation with the expression of $g(u_3)$ deduced from equation [A3.19]:

$$\left(-g(v)+\frac{u_3}{g(v)}\cdot\frac{\partial g(v)}{\partial v}\right)\cdot g(u_3)+g(v)=0 \qquad [\text{A2.29}]$$

By dividing expression [A3.29] by $g(v)$, it becomes:

$$\left(-1+\frac{u_3}{(g(v))^2}\cdot\frac{\partial g(v)}{\partial v}\right)\cdot g(u_3)+1=0 \qquad [\text{A3.30}]$$

And by dividing expression [A3.30] by $u_3.g(u_3)$, we get:

$$\frac{g(u_3)-1}{u_3.g(u_3)}=\frac{1}{(g(v))^2}\cdot\frac{\partial g(v)}{\partial v} \qquad [\text{A3.31}]$$

For the equality between these two independent functions to be satisfied, they must be equal to a constant denoted by A, which thus gives:

$$\frac{1}{(g(v))^2}\cdot\frac{\partial g(v)}{\partial v}=A \qquad [\text{A3.32}]$$

and:

$$\frac{g(u_3)-1}{u_3.g(u_3)}=A \qquad [\text{A3.33}]$$

The integration of equation [A3.32] leads to:

$$g(v)=(\alpha+\beta.v)^{-1} \qquad [\text{A3.34}]$$

where $\beta=-A$ and α is an integration constant.

The resolution of equation [A3.33] leads to:

$$g(u_3)=(1-A.u_3)^{-1}=(1+\beta.u_3)^{-1} \qquad [\text{A3.35}]$$

The equality of expressions [A3.34] and [A3.35] gives $\alpha = 1$. By combining equations [A3.19] and [A3.34], we can deduce that:

$$g(v) = (1 + \beta.v)^{-1} = \frac{1}{\phi(v)} \cdot \frac{\partial \phi(v)}{\partial v} \qquad [A3.36]$$

Hence:

$$(1 + \beta.u) \cdot \left(\frac{\partial \varphi}{\partial v} \right) - \phi(u) = 0 \qquad [A3.37]$$

In order to determine the required form of the reference-invariant standard dimensionless material function ϕ, the solution of this differential equation is obtained by integration, to within one constant. This constant can be determined by using one of the properties of the standard dimensionless material function: the slope of its tangent at the reference abscissa ($u=0$) is equal to 1; therefore, the derivative of ϕ, denoted by ϕ', has a value of 1 in $u=0$: $\phi'(0)=1$.

We can, therefore, show that equation [A3.37] has the solution of:

$$\begin{cases} \beta \neq 0 \rightarrow \phi(u) = (1 + \beta.u)^{\frac{1}{\beta}} \\ \beta = 0 \rightarrow \phi(u) = \exp(u) \end{cases} \qquad [A3.38]$$

Consequently, a standard dimensionless material function is reference-invariant if and only if it has the form of these two equations.

Appendix 4

Cases Where the Analytical Expression of the Material Function Is Known

In this appendix, we will apply the guidelines given in section 4.4 of Chapter 4 to unambiguously establish the complete space of dimensionless numbers influencing the target internal measure for a process involving a *Bingham or a Williamson–Cross fluid*. In both cases, there is a known analytical expression of the dimensional material function: the rheological law.

A4.1. Bingham fluid

We have seen that, for a Bingham fluid, the dimensional material function describing the variation in apparent viscosity μ_a with the shear rate $\dot{\gamma}$ is expressed by:

$$\mu_a(\dot{\gamma}) = \frac{\tau_y}{\dot{\gamma}} + \mu_p \qquad [A4.1]$$

where τ_y is the yield stress and μ_p is the plastic viscosity of the fluid when $\tau > \tau_y$.

It does not have the form of a reference-invariant function (see equations [4.91] or [4.92]). The reference shear rate $\dot{\gamma}_0$ must, therefore, be added to the initial relevance list of physical quantities influencing the target variable for a process involving a Bingham fluid.

According to equation [4.103], the dimensionless number to introduce into the initial relevance list is $a_0.\dot{\gamma}_0$. In this example, $a_0.\dot{\gamma}_0$ can be expressed analytically. By definition, we have:

$$\begin{cases} a_0.\dot{\gamma}_0 = \dfrac{1}{\mu_0}\left(\dfrac{\mathrm{d}\mu_a}{\mathrm{d}\dot{\gamma}}\right)_{\dot{\gamma}=\dot{\gamma}_0}.\dot{\gamma}_0 \\[2mm] u = a_0.\dot{\gamma}_0\left[\dfrac{\dot{\gamma}}{\dot{\gamma}_0}-1\right] \end{cases} \qquad [A4.2]$$

Given that:

$$\begin{cases} \mu_0 = \mu_a\left(\dot{\gamma}_0\right) = \dfrac{\tau_y}{\dot{\gamma}_0}+\mu_p \\[2mm] \left(\dfrac{\mathrm{d}\mu_a}{\mathrm{d}\dot{\gamma}}\right)_{\dot{\gamma}=\dot{\gamma}_0} = -\dfrac{\tau_y}{\dot{\gamma}_0^{\,2}} \end{cases} \qquad [A4.3]$$

We can deduce:

$$a_0 = \frac{1}{\mu_0}\left(-\frac{\tau_y}{\dot{\gamma}_0^{\,2}}\right) \qquad [A4.4]$$

Let us now introduce the dimensionless number Bi defined as:

$$Bi = \frac{\tau_y}{\mu_0.\dot{\gamma}_0} \qquad [A4.5]$$

By combining equations [A4.4] and [A4.5], we get:

$$\begin{cases} a_0.\dot{\gamma}_0 = (-Bi.\dot{\gamma}_0^{-1}).\dot{\gamma}_0 = -Bi \\ u = -Bi.\left[\dfrac{\dot{\gamma}}{\dot{\gamma}_0} - 1\right] = Bi.\left[1 - \dfrac{\dot{\gamma}}{\dot{\gamma}_0}\right] \end{cases}$$ [A4.6]

Therefore, the analytical expression of the dimensionless number $a_0.\dot{\gamma}_0$ produces a dimensionless number, Bi, which is the Bingham number (see footnote no. 5 in Chapter 4). Therefore, in order to define the configuration of the material, it would seem appropriate to replace $a_0.\dot{\gamma}_0$ by Bi, thus giving:

$$\{\pi_m\} = \{a_0.\dot{\gamma}_0\} = \{Bi\}$$ [A4.7]

In summary, the initial relevance list of physical quantities influencing the target variable will be supplemented, in the case of a process involving a Bingham fluid, by:

$$\{\dot{\gamma}_0, \mu_0, Bi\}$$ [A4.8]

COMMENTS.–

– Not all the rheological parameters describing the dimensional material function of a Bingham fluid can be found in the set of internal measures $\{\pi_m\}$. This is the case for plastic viscosity μ_p.

– It is possible to determine the analytical expression of the standard dimensionless material function w associated with a Bingham fluid depending on the argument u.

From equation [A4.6], we can deduce that:

$$\dot{\gamma} = \dot{\gamma}_0.\left[1 - \frac{u}{Bi}\right]$$ [A4.9]

By inserting equation [A4.9] into the expression defining the standard dimensionless material function w, we obtain:

$$w(u) = \frac{\mu_a}{\mu_0} = \frac{1}{\mu_0} \cdot \left[\frac{\tau_y}{\dot{\gamma}_0 . (1 - \frac{u}{Bi})} + \mu_p \right] = \frac{\tau_y}{\dot{\gamma}_0 . \mu_0 . (1 - \frac{u}{Bi})} + \frac{\mu_p}{\mu_0}$$

$$= \frac{Bi}{(1 - \frac{u}{Bi})} + \frac{\mu_0 - \frac{\tau_y}{\dot{\gamma}_0}}{\mu_0} = \frac{Bi}{(1 - \frac{u}{Bi})} + 1 - Bi$$

[A4.10]

$$= \frac{Bi + (1 - Bi) . (1 - \frac{u}{Bi})}{(1 - \frac{u}{Bi})}$$

$$w(u) = 1 + \frac{u}{(1 - \frac{u}{Bi})}$$

Equation [A4.10] confirms that the standard dimensionless material function $w(u)$ associated with a Bingham fluid does not have the form of a reference-invariant function (equations [4.77] or [4.78]). Moreover, it can be observed that the dimensionless number which appears in $w(u)$ is indeed the one previously identified in $\{a_0 . \dot{\gamma}_0\}$ (equation [A4.6]), namely Bi.

A4.2. Williamson–Cross fluid

For a Williamson–Cross fluid, the dimensional material function describing the variation of the apparent viscosity with the shear rate involves three parameters and is expressed by:

$$\mu_a(\dot{\gamma}) = \frac{\mu_w}{1 + (t_w . \dot{\gamma})^{1-n}}$$

[A4.11]

where n is the flow index, μ_w is the parameter that describes the Newtonian behavior of the fluid at low shear rates and t_w is the characteristic time of the transition between the Newtonian and shear-thinning behavior.

It does not have the form of a reference-invariant function (see equations [4.91] or [4.92]). The reference shear rate $\dot{\gamma}_0$ must, therefore, be added to the initial relevance list of physical quantities influencing the target variable for a process involving a Williamson–Cross fluid.

According to equation [4.103], the dimensionless number to introduce into the initial relevance list is $\{a_0 . \dot{\gamma}_0\}$. In this example also, $\{a_0 . \dot{\gamma}_0\}$ can be expressed analytically. By using equation [A4.2] and given that:

$$\begin{cases} \mu_0 = \mu_a(\dot{\gamma}_0) = \dfrac{\mu_w}{1+\left(t_w . \dot{\gamma}_0\right)^{1-n}} \\[4mm] \left(\dfrac{d\mu_a}{d\dot{\gamma}}\right)_{\dot{\gamma}=\dot{\gamma}_0} = \dfrac{\mu_w . t_w . (n-1).\left(t_w . \dot{\gamma}_0\right)^{-n}}{\left[1+\left(t_w . \dot{\gamma}_0\right)^{1-n}\right]^2} \end{cases} \qquad [A4.12]$$

It can be deduced that:

$$\begin{cases} a_0 = \dfrac{1}{\mu_0} . \left[\dfrac{\mu_w . t_w . (n-1).\left(t_w . \dot{\gamma}_0\right)^{-n}}{\left[1+\left(t_w . \dot{\gamma}_0\right)^{1-n}\right]^2}\right] = \dfrac{1}{\dfrac{\mu_w}{1+\left(t_w . \dot{\gamma}_0\right)^{1-n}}} . \left[\dfrac{\mu_w . t_w . (n-1).\left(t_w . \dot{\gamma}_0\right)^{-n}}{\left[1+\left(t_w . \dot{\gamma}_0\right)^{1-n}\right]^2}\right] \\[6mm] a_0 = \dfrac{t_w . (n-1).\left(t_w . \dot{\gamma}_0\right)^{-n}}{1+\left(t_w . \dot{\gamma}_0\right)^{1-n}} = \dfrac{(n-1).\left(t_w . \dot{\gamma}_0\right)^{-n}}{1+\left(t_w . \dot{\gamma}_0\right)^{1-n}} . \dot{\gamma}_0^{-1} \end{cases} \qquad [A4.13]$$

Hence:

$$\begin{cases} a_0 . \dot{\gamma}_0 = \dfrac{(n-1).\left(t_w . \dot{\gamma}_0\right)^{1-n}}{1+\left(t_w . \dot{\gamma}_0\right)^{1-n}} \\[6mm] u = a_0 . \dot{\gamma}_0 . \left[\dfrac{\dot{\gamma}}{\dot{\gamma}_0}-1\right] = \dfrac{(n-1).\left(t_w . \dot{\gamma}_0\right)^{1-n}}{1+\left(t_w . \dot{\gamma}_0\right)^{1-n}} . \left[\dfrac{\dot{\gamma}}{\dot{\gamma}_0}-1\right] \end{cases} \qquad [A4.14]$$

Therefore, the analytical expression of the dimensionless number $\{a_0 . \dot{\gamma}_0\}$ produces two dimensionless numbers:

– n, which is the flow index of the fluid;

$-t_w \cdot \dot{\gamma}_0$, which can be seen as an internal measure of the characteristic time of transition between the Newtonian and shear-thinning behavior.

Since these dimensionless numbers are representative indicators of the shear-thinning behavior of the fluid, it is advisable to consider them in the configuration of the material. This, therefore, gives:

$$\{\pi_m\} = \{a_0 \cdot \dot{\gamma}_0\} = \{t_w \cdot \dot{\gamma}_0, n\}$$
[A4.15]

In the end, the initial relevance list of physical quantities influencing the target variable will be supplemented, in the case of a process involving a Williamson–Cross fluid, by:

$$\{\dot{\gamma}_0, \mu_0, t_w \cdot \dot{\gamma}_0, n\}$$
[A4.16]

COMMENTS.–

– Once again, not all the rheological parameters describing the dimensional material function of a Williamson–Cross fluid can be found in the set of internal measures $\{\pi_m\}$; this is the case for μ_w.

– It is possible to determine the analytical expression of the standard dimensionless material function w associated with a Williamson–Cross fluid depending on the argument u. From equation [A4.14], it can be deduced that:

$$
\begin{aligned}
u &= \dot{\gamma}_0 + \frac{\dot{\gamma}_0 \cdot u \cdot [1 + (t_w \cdot \dot{\gamma}_0)^{1-n}]}{(n-1) \cdot (t_w \cdot \dot{\gamma}_0)^{1-n}} \\
&= \dot{\gamma}_0 \cdot \left[1 + \frac{u \cdot [1 + (t_w \cdot \dot{\gamma}_0)^{1-n}]}{(n-1) \cdot (t_w \cdot \dot{\gamma}_0)^{1-n}} \right]
\end{aligned}
$$
[A4.17]

The standard dimensionless material function associated with a Williamson–Cross fluid is defined by:

$$w(u) = \frac{\mu_a(\dot{\gamma})}{\mu_0} = \frac{1}{\mu_0} \cdot \frac{\mu_w}{1 + (t_w \cdot \dot{\gamma})^{1-n}}$$
[A4.18]

By inserting equation [A4.17] into equation [A4.18], we have:

$$w(u) = \frac{\mu_a(\dot{\gamma})}{\mu_0} = \frac{1}{\mu_0} \cdot \frac{\mu_w}{1 + \left(t_w \cdot \dot{\gamma}_0 \cdot \left[1 + \dfrac{u \cdot \left[1 + \left(t_w \cdot \dot{\gamma}_0 \right)^{1-n} \right]}{(n-1) \cdot \left(t_w \cdot \dot{\gamma}_0 \right)^{1-n}} \right] \right)^{1-n}}$$

$$= \frac{1 + \left(t_w \cdot \dot{\gamma}_0 \right)^{1-n}}{\mu_w} \cdot \frac{\mu_w}{1 + \left(t_w \cdot \dot{\gamma}_0 \right)^{1-n} \left(1 + \dfrac{u \cdot \left[1 + \left(t_w \cdot \dot{\gamma}_0 \right)^{1-n} \right]}{(n-1) \cdot \left(t_w \cdot \dot{\gamma}_0 \right)^{1-n}} \right)^{1-n}} \qquad \text{[A4.19]}$$

$$= \frac{1 + \left(t_w \cdot \dot{\gamma}_0 \right)^{1-n}}{1 + \left(t_w \cdot \dot{\gamma}_0 \right)^{1-n} \cdot \left(1 + \dfrac{u}{(n-1)} \cdot \left[1 + \dfrac{1}{\left(t_w \cdot \dot{\gamma}_0 \right)^{1-n}} \right] \right)^{1-n}}$$

Equation [A4.19] confirms that the standard dimensionless material function $w(u)$ associated with a Williamson–Cross fluid does not have the form of a reference-invariant function (equations [4.77] or [4.78]). Moreover, it is readily seen that the dimensionless numbers which appear in $w(u)$ are indeed the same as those previously identified in $a_0 \cdot \dot{\gamma}_0$ (equation [A4.14]), namely $t_w \cdot \dot{\gamma}_0$ and n.

Appendix 5

Cases Where There Is No Known Analytical Expression of the Material Function

In this appendix, we will apply the guidelines provided in Chapter 4 in order to unambiguously establish the complete space of dimensionless numbers influencing the target internal measure for a process involving an *aqueous mixture whose surface tension varies depending on the volume fraction of butanol*[1]. Just as in example 4.5 of Chapter 4 (dependence of the viscosity of Newtonian fluids on temperature), there is no known analytical expression of the material function.

The physical property being considered here is the surface tension of the aqueous mixture (denoted by σ) and the variable the volume fraction of butanol (denoted by x). The form of the dimensional material function is, therefore, $\sigma(x)$.

Experimental measurements [LOU 02] helped to determine, for different values of volume fractions of butanol, the surface tension of the aqueous mixture (at 20°C). The set of discrete points (x_i, σ_i) obtained is graphically represented in Figure A5.1(a).

---

1 We chose this example, which is more marginal than that of the dependence of the Newtonian viscosity on temperature, to show that the theoretical framework and the tools defined in Chapter 4 apply to any variable physical property.

Since we do not have an analytical expression to describe $\sigma(x)$, it is necessary to find with which mathematical function these points can be approximated.

The same method as the one described in example 4.5 in Chapter 4 must be applied. However, it is impossible to find a *single* approximated function able to describe the variations of σ in the entire range of volume fractions x. In order to overcome this problem, a simple solution consists of *finding an approximated function by intervals*. Eventually, this gives[2]:

– for $0 \le x \le 0.1$:

$$\sigma = -3.988 \times 10^5 \cdot x^5 + 1.162 \times 10^4 \cdot x^4 - 1.281 \times 10^{3} \cdot x^3 + 6.901 \times 10^2 \cdot x^2 - 2.168 \times 10^1 \cdot x + 7.220 \times 10^{-1} \qquad \text{[A5.1]}$$

– for $0.1 \le x \le 1$:

$$\sigma = -1.031 \times 10^{-2} \cdot x + 0.272 \qquad \text{[A5.2]}$$

The adequacy of these two approximated functions with the discrete points is shown in Figure A5.1(a).

It can be deduced from this that:

– for $0 \le x \le 0.1$:

$$\frac{d\sigma}{dx} = -5 \times 3.988 \times 10^5 \cdot x^4 + 4 \times 1.162 \times 10^4 \cdot x^3 - 3 \times 1.281 \times 10^{3} \cdot x^2 + 2 \times 6.901 \times 10^2 \cdot x - 2.168 \times 10^1 \qquad \text{[A5.3]}$$

– for $0.1 \le x \le 1$:

$$\frac{d\sigma}{dx} = -1.031 \cdot 10^{-2} \qquad \text{[A5.4]}$$

Let us choose a reference volume fraction $x_0 \ne 0$, for example $x_0 = 0.02$. It gives:

---

2 In these relationships (and in those which result from them), surface tension is expressed in N/m and the volume fraction of butanol is dimensionless.

$$\sigma_0 = \sigma(x_0) = 0.479 \ \text{N.m}^{-1} \tag{A5.5}$$

$$\left(\frac{d\sigma}{dx}\right)_{x=x_0} = -5 \times 3.988 \times 10^5 \cdot x_0^4 + 4 \times 1.162 \times 10^4 \cdot x_0^3 - 3 \times 1.281 \times 10^{3} \cdot$$
$$x_0^2 + 2 \times 6.901 \times 10^2 \cdot x_0 - 2.168 \times 10^1 = -6.066 \ \text{N.m}^{-1} \tag{A5.6}$$

$$a_0 = \frac{1}{\sigma_0}\left(\frac{d\sigma}{dx}\right)_{x=x_0} = -12.637 \tag{A5.7}$$

It is possible to deduce from the argument u of the standard dimensionless material function that:

$$u = a_0 \cdot x_0 \cdot \left[\frac{x}{x_0} - 1\right] \tag{A5.8}$$

The standard dimensionless material function w can be calculated for each interval as follows:

– for $0 \leq x \leq 0.1$:

$$w(u) = \frac{-3.988 \times 10^5 \cdot x^5 + 1.162 \times 10^4 \cdot x^4 - 1.281 \times 10^{3} \cdot x^3}{0.479} + \frac{6.901 \times 10^2 \cdot x^2 - 2.168 \times 10^1 \cdot x + 7.220 \times 10^{-1}}{0.479} \tag{A5.9}$$

– for $0.1 \leq x \leq 1$:

$$w(u) = \frac{-1.031 \times 10^{-2} \cdot x + 0.272}{0.479} \tag{A5.10}$$

Figure A5.1(b) shows the graphical representation of the standard dimensionless material function w according to its argument u in the form of a set of discrete points (u_i, w_i). The two conditions issued from the standardization, namely $w(u = 0) = 1$ and $w'(u = 0) = 1$, have been properly satisfied.

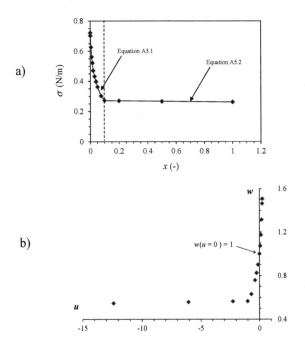

Figure A5.1. *Variation of the surface tension of aqueous mixtures according to the volume fraction of butanol: a) dimensional material function (unbroken line: equations [A5.1] and [A5.2]) and b) standard dimensionless material function (the reference volume fraction chosen is $x_0 = 0.02$)*

To know whether the set of discrete points (u_i, w_i) given in Figure A5.1(b) corresponds to a reference-invariant standard dimensionless material function, we must examine whether they can be approximated by a mathematical function of the form of equation [4.77] or [4.78]. It can be demonstatred that none of these two functions makes it possible to approximate, with an acceptable margin for error, the set of discrete points (u_i, w_i). Therefore, the approximated function of the material function associated with the dependence of the surface tension of aqueous mixtures on the volume fraction of butanol does not have invariance properties.

What are the consequences of these results on the implementation of dimensional analysis in a process involving this material function?

According to the discussion in Chapter 4, the initial relevance list of physical quantities influencing the target variable required must be supplemented by:

$$\{x_0 ; \sigma_0 ; \{\pi_m\}\}$$

[A5.11]

The reference volume fraction x_0 must be added to the relevance list when the material function does not have invariance properties.

According to equation [4.103], the dimensionless number $\{\pi_m\} = \{a_0 \cdot x_0\}$ appearing in the argument u of the standard dimensionless material function must be added to the initial relevance list. In this case, it is calculated using equations [A5.5] and [A5.7] whereby:

$$a_0 \cdot x_0 = \frac{1}{\sigma_0} \cdot \left(\frac{d\sigma}{dx}\right)_{x=x_0} \cdot x_0 = -0.253$$

[A5.12]

In the end, the initial relevance list of physical quantities influencing the target variable required, for a process involving an aqueous mixture whose surface tension varies with the volume fraction of butanol, will be supplemented by:

$$\{x_0 ; \sigma_0 ; a_0 \cdot x_0\}$$

[A5.13]

Appendix 6

Relevant Choice of the Reference Abscissa for Non-Newtonian Fluids

This appendix is a supplement to section 4.4.3 of Chapter 4. The objective is to revisit the examples of Bingham and Williamson–Cross fluids to show that certain choices of reference abscissa are more relevant than others. Also in this appendix, we will suggest *summary tables* which contain, for various non-Newtonian fluids, the expressions of dimensional and standard dimensionless material functions, dimensionless numbers $\{\pi_m\}$ and a relevant choice of reference abscissa.

A6.1. Bingham fluid

We have seen (Appendix 4) that, for a process involving a Bingham fluid, the initial relevance list of physical quantities influencing the target variable must be supplemented by:

$$\{\dot{\gamma}_0, \mu_0, Bi\} \tag{A6.1}$$

We will show that it is relevant to choose a reference shear rate $\dot{\gamma}_0$ so that:

$$\dot{\gamma}_0 = \frac{\tau_y}{\mu_p} \tag{A6.2}$$

where τ_y is the yield stress and μ_p is the plastic viscosity of the fluid when $\tau > \tau_y$.

In this case, the apparent viscosity calculated at the reference shear rate, μ_0, is:

$$\mu_0 = \frac{\tau_y}{\dot{\gamma}_0} + \mu_p = \frac{\tau_y}{\dfrac{\tau_y}{\mu_p}} + \mu_p = 2\mu_p \qquad\qquad [A6.3]$$

The dimensionless number Bi defined by equation [4.116] becomes:

$$Bi = \frac{\tau_y}{\mu_0 \cdot \dot{\gamma}_0} = \frac{\tau_y}{\left(2\mu_p\right)\cdot\left(\dfrac{\tau_y}{\mu_p}\right)} = \frac{1}{2} \qquad\qquad [A6.4]$$

Due to this choice of reference shear rate, the dimensionless number Bi becomes equal to a constant (1/2).

As a result, if the chosen reference shear rate $\dot{\gamma}_0$ is that given by equation [A6.2], the relevance list of physical quantities influencing the target variable will only be supplemented, in the case of a process involving a Bingham fluid, by:

$$\left\{\dot{\gamma}_0 = \frac{\tau_y}{\mu_p},\ \mu_0 = 2\mu_p\right\} \qquad\qquad [A6.5]$$

Equation [A6.5] includes one internal measure less, Bi, in comparison to equation [A6.1] obtained for any reference shear rate $\dot{\gamma}_0$. Nevertheless, it is important to remember that even if the internal measure Bi no longer appears explicitly in equation [A6.5], it is still part of the configuration of the system (star graph), but its value is fixed and is equal to ½.

A6.2. Williamson–Cross fluid

We have demonstrated (Appendix 4) that, for a process involving a Williamson–Cross fluid, the initial relevance list of physical quantities influencing the target variables must be supplemented by:

$$\left\{\dot{\gamma}_0,\ \mu_0,\ t_w\cdot\dot{\gamma}_0,\ n\right\} \qquad\qquad [A6.6]$$

We will now show that it is relevant to choose a reference shear rate $\dot{\gamma}_0$ so that:

$$\dot{\gamma}_0 = \frac{1}{t_w} \qquad\qquad [A6.7]$$

where t_w is the characteristic time of the transition between the Newtonian and shear-thinning behavior.

Therefore, the apparent viscosity calculated at the reference shear rate, μ_0, is:

$$\mu_0 = \frac{\mu_w}{1+\left(t_w \cdot \dot{\gamma}_0\right)^{1-n}} = \frac{\mu_w}{1+\left(t_w \cdot \dfrac{1}{t_w}\right)^{1-n}} = \frac{\mu_w}{2} \qquad\qquad [A6.8]$$

The dimensionless number $t_w \cdot \dot{\gamma}_0$ then becomes:

$$t_w \cdot \dot{\gamma}_0 = t_w \cdot \frac{1}{t_w} = 1 \qquad\qquad [A6.9]$$

Due to this choice of reference shear rate, the dimensionless number $t_w \cdot \dot{\gamma}_0$ becomes equal to a constant (1).

As a result, if the chosen reference shear rate $\dot{\gamma}_0$ is that given by equation [A6.7], the relevance list of physical quantities influencing the target variable will only be supplemented, in the case of a process involving a Williamson–Cross fluid, by:

$$\left\{\dot{\gamma}_0 = \frac{1}{t_w}, \ \mu_0 = \frac{\mu_w}{2}, \ n\right\} \qquad\qquad [A6.10]$$

Equation [A6.10] includes one less internal measure, $t_w \cdot \dot{\gamma}_0$, in comparison with equation [A6.6] obtained for any reference shear rate $\dot{\gamma}_0$. However, it is important to remember that even if the dimensionless number

$t_w \cdot \dot{\gamma}_0$ no longer appears explicitly in equation [A6.10], it is still part of the configuration of the system (star graph), but its value is fixed and is equal to 1.

A6.3. Summary for non-Newtonian fluids

In Chapter 4, we defined the actions to be taken in order to unambiguously establish the complete space of dimensionless numbers influencing the target internal measure in the case of a process involving a material with a variable physical property. We illustrated these actions using several examples where the physical property was the apparent viscosity and the variable the shear rate. Since processes involving non-Newtonian fluids are common, we will summarize below the results obtained in Chapter 4 and in Appendices 4–6 in the form of tables.

Type of fluid	Dimensional material function $\mu_a(\dot{\gamma})$	Argument u
Pseudoplastic	$k \cdot \dot{\gamma}^{n-1}$	$\left(\dfrac{\dot{\gamma}-\dot{\gamma}_0}{\dot{\gamma}_0}\right) \cdot (n-1)$
Bingham	$\dfrac{\tau_y}{\dot{\gamma}} + \mu_p$	$Bi.\left[1-\dfrac{\dot{\gamma}}{\dot{\gamma}_0}\right]$
Herschel–Bulkley	$\dfrac{\tau_y}{\dot{\gamma}} + k \cdot \dot{\gamma}^{n-1}$	$\left(-Bi+(n-1).[1-Bi]\right).\left[\dfrac{\dot{\gamma}}{\dot{\gamma}_0}-1\right]$
Williamson–Cross	$\dfrac{\mu_w}{1+\left(t_w \cdot \dot{\gamma}\right)^{1-n}}$	$\dfrac{(n-1)\left(t_w \cdot \dot{\gamma}_0\right)^{1-n}}{1+\left(t_w \cdot \dot{\gamma}_0\right)^{1-n}} \cdot \left[\dfrac{\dot{\gamma}}{\dot{\gamma}_0}-1\right]$

Table A6.1. *Expressions for common non-Newtonian fluid models of the dimensional material function and the argument u associated with the standard dimensionless material function*

Type of fluid	Standard dimensionless material function $w(u)$
Pseudoplastic	$$\left(\frac{u}{n-1}+1\right)^{n-1}$$
Bingham	$$1+\frac{u}{(1-\dfrac{u}{Bi})}$$
Herschel–Bulkley	$$\frac{Bi}{\dfrac{u}{n.(1-Bi)-1}+1}+(1-Bi).\left[\frac{u}{n.(1-Bi)-1}+1\right]^{n-1}$$
Williamson–Cross	$$\frac{1+\left(t_w.\dot{\gamma}_0\right)^{1-n}}{1+\left(t_w.\dot{\gamma}_0\right)^{1-n}.\left(1+\dfrac{u}{(n-1)}.\left[1+\dfrac{1}{\left(t_w.\dot{\gamma}_0\right)^{1-n}}\right]\right)^{1-n}}$$

Table A6.2. *Expressions for common non-Newtonian fluid models of the standard dimensionless material function w(u)*

Type of fluid	Physical quantities / supplementary dimensionless numbers to introduce to the initial relevance list	Relevant choice of $\dot{\gamma}_0$
Pseudoplastic	$\{\ \mu_0\ ;\ n\ \}$	Not listed
Bingham	$\left\{ \dot{\gamma}_0, \mu_0, Bi = \dfrac{\tau_y}{\mu_0 \cdot \dot{\gamma}_0} \right\}$	$\dot{\gamma}_0 = \dfrac{\tau_y}{\mu_p}$ hence $\left\{ \dot{\gamma}_0 = \dfrac{\tau_y}{\mu_p},\ \mu_0 = 2\mu_p \right\}$ with $Bi = 1/2$
Herschel–Bulkley	$\{\ \dot{\gamma}_0\ ;\ \mu_0\ ;\ n\ ;\ Bi\ \}$	$\dot{\gamma}_0 = \left(\dfrac{\tau_y}{k} \right)^{1/n}$ $\left\{ \dot{\gamma}_0 = \left(\dfrac{\tau_y}{k} \right)^{1/n} \begin{array}{c} \text{hence} \\ \ \end{array},\ \mu_0 = 2k^{1/n} \cdot (\tau_y)^{\frac{n-1}{n}},\ n \right\}$ with $Bi = 1/2$
Williamson–Cross	$\{\ \dot{\gamma}_0\ ;\ \mu_0\ ;\ n\ ;\ t_w \cdot \dot{\gamma}_0\ \}$	$\dot{\gamma}_0 = \dfrac{1}{t_w}$ hence $\left\{ \dot{\gamma}_0 = \dfrac{1}{t_w},\ \mu_0 = \dfrac{\mu_w}{2},\ n \right\}$ with $t_w \cdot \dot{\gamma}_0 = 1$

Table A6.3. *Physical quantities/dimensionless numbers to introduce to the initial relevance list and the relevant choice of the reference shear rate for common non-Newtonian fluid models*

Bibliography

[AND 12] ANDRÉ C., DEMEYRE J.F., GATUMEL C. et al., "Dimensional analysis of a planetary mixer for homogenizing of free flowing powders: mixing time and power consumption", *Chemical Engineering Journal*, vol. 198–199, pp. 371–378, 2012.

[BAL 07] BALERIN C., AYMARD P., DUCEPT F. et al., "Effect of formulation and processing factors on the properties of liquid food foams", *Journal of Food Engineering*, vol. 78, pp. 802–809, 2007.

[BEC 76] BECKER H.A., *Dimensionless Parameters: Theory and Methodology*, Applied Science Publishers Ltd., London, 1976.

[BLA 12] BLANPAIN-AVET P., HÉDOUX A., GUINET Y. et al., "Analysis by Raman spectroscopy of the conformational structure of whey proteins constituting fouling deposits during the processing in a heat exchanger", *Journal of Food Engineering*, vol. 110, pp. 86–94, 2012.

[BOR 02] CYRULNIK B., *Un merveilleux malheur*, Odile Jacob, 2002.

[BRO 70] BROWN D.E., PITT K., "Drop breakup in a stirred liquid–liquid contactor", *Proceedings Chemeca*, Melbourne, Sydney, p. 83, 1970.

[BUC 21] BUCKINGHAM E., "On plastic flow through capillary tubes", *ASTM Proceedings*, vol. 21, pp. 1154–1156, 1921.

[CAL 58] CALDERBANK P.H., "Physical rate processes in industrial fermentation, part I: the interfacial area in gas–liquid contacting with mechanical agitation", *Transactions of the Institution of Chemical Engineers*, vol. 36, pp. 443–463, 1958.

[CRA 75] CRANK J., *The Mathematics of Diffusion*, Oxford Science Publications, Oxford, 1975.

[CUS 97] CUSSLER E.L., *Diffusion – Mass Transfer in Fluid Systems*, 3rd ed., Cambridge University Press, 1997.

[DEL 00] DELAPLACE G., MOREAU A., BELAUBRE N. *et al.*, "Effet de la position du mobile par rapport au fond de cuve sur la consommation de puissance de différents agitateurs plans verticaux", *Proceedings du 5ème Colloque Prosetia*, Toulouse, France, pp. 109–114, 1st–2nd February 2000.

[DEL 04] DELAPLACE G., GUÉRIN R., MOREAU A. *et al.*, "Experimental and CFD study of power consumption with gate agitators – effects of bottom clearance and scale up", in BOBEANU C. (ed.), *2nd International Conference on Simulation in Food and Bio Industries, Foodsim,* Wageningen, The Netherlands, also in, *Proceedings of the European Simulation and Modelling Conference*, Paris, pp. 338–344, 16–18 June 2004.

[DEL 05] DELAPLACE G., GUERIN R., LEULIET J.C., "Dimensional analysis for planetary mixer: modified power and Reynolds numbers", *AIChE Journal*, vol. 51, pp. 1–7, 2005.

[DEL 07] DELAPLACE G., THAKUR R.K., BOUVIER L. *et al.*, "Dimensional analysis for planetary mixer: mixing time and Reynolds numbers", *Chemical Engineering Science*, vol. 62, pp. 1442–1447, 2007.

[DEL 08] DELAPLACE G., THAKUR R.K., BOUVIER L. *et al.*, "Influence of rheological behavior of purely viscous fluids on analytical residence time distribution in straight tubes", *Chemical Engineering Technology*, vol. 31, pp. 231–236, 2008.

[DEL 09] DELAPLACE G., MAINGONNAT J.-F., ZAÏD I. *et al.*, "Scale up of ohmic heating processing with food material", in VOROBIEV E., LEBOVKA N., VAN HECKE E. *et al.* (eds), *Proceedings of the International Conference on Bio & Food Electrotechnologies (BFE2009)*, Compiègne, pp. 101–107, 22–23 October 2009.

[FRY 06] FRYER P.J., CHRISTIAN G.K., LIU W., "How hygiene happens: physics and chemistry of cleaning", *International Journal of Dairy Technology*, vol. 59, pp. 76–84, 2006.

[GAR 09] GARCIA-OCHOA F., GOMEZ E., "Bioreactor scale-up and oxygen transfer rate in microbial processes: an overview", *Biotechnology Advances*, vol. 27, pp. 153–176, 2009.

[GHN 08a] GHNIMI S., Etude des performances thermique et hydraulique d'une cellule à effet Joule direct avec Jet de fluide: applications aux fluides visqueux et encrassants, Thesis, University of Technology of Compiègne, 2008

[GHN 08b] GHNIMI S., MALASPINA N.F., DELAPLACE G. *et al.*, "Design and performance evaluation of a new ohmic heating unit for thermal processing of highly viscous liquids", *Chemical Engineering Research and Design*, vol. 86, pp. 626–632, 2008

[GHN 09] GHNIMI S., FLACH-MALASPINA N., MAINGONNAT J.F. *et al.*, "Axial temperature profile of ohmically heated fluid jet: analytical model and experimental validation", *Chemical Engineering Science*, vol. 64, no. 13, pp. 3188–3196, 1 July 2009.

[GOV 87] GOVIER G.W., AZIZ K., *The Flow of Complex Mixtures in Pipes*, Robert E. Krieger Publishing Co, Huntington, New York, 1987.

[GRI 04] GRIJSPEERDT K., MORTIER L., DE BLOCK J. *et al.*, "Applications of modeling to optimize ultra-high temperature milk heat exchangers with respect to fouling", *Food Control*, vol. 15, pp. 117–130, 2004.

[GUR 23] GURNEY H.P., LURIE J., "Charts for estimating temperature distributions in heating or cooling solid shapes", *Ind. Eng. Chem.*, vol. 15, pp. 1170–1172, 1923.

[HAS 12] HASSAN R., LOUBIÈRE K., LEGRAND J. *et al.*, "A consistent dimensional analysis of gas-liquid mass transfer in an aerated stirred tank containing purely viscous fluids with shear-thinning properties", *Chemical Engineering Journal*, vol. 184, pp. 42–56, 2012.

[HED 52] HEDSTRÖM B.O.A., "Flow of plastic materials in pipes", *Journal of Industrial and Engineering Chemistry*, vol. 44, pp. 651–656, 1952.

[JEA 10] JEANTET R., SCHUCK P., SIX T. *et al.*, "Stirring speed, temperature and solid concentration influence on micellar casein powder rehydration time", *Dairy Science and Technology*, vol. 90, pp. 225–236, 2010.

[JOH 57] JOHNSTONE R.E., THRING M.W., *Pilot Plants, Model and Scale-up Methods in Chemical Engineering*, McGraw-Hill Inc., Columbus, 1957.

[JOS 82] JOSHI J.B., PANDIT A.B., SHARMA M.M., "Mechanically agitated gas-liquid reactors", *Chemical Engineering Science*, vol. 376, pp. 813–844, 1982.

[JUD 82] JUDAT H., "Gas/liquid mass transfer in stirred vessels – a critical review", *German Chemical Engineering*, vol. 5, pp. 357–363, 1982.

[JUN 05] JUN S., PURI V.M., "Fouling models for heat exchangers: a review", *Journal of Food Process Engineering*, vol. 28, pp. 1–34, 2005.

[LÁS 64] LÁSZLO A., "Systematization of dimensionless quantities by group theory", *International Journal of Heat and Mass Transfer*, vol. 7, pp. 423–430, 1964.

[LEW 24] LEWIS W.K., WHITMAN W.G., "Principle of gas absorption", *Journal of Industrial and Engineering Chemistry*, vol. 16, pp. 1215–1220, 1924.

[LOU 02] LOUBIÈRE K., Croissance et détachement de bulles générées par des orifices rigides et flexibles dans des phases liquides Newtoniennes: Etude expérimentale et modélisation, Thesis no. 663, Institut National des Sciences Appliqués, Toulouse, 2002.

[MAR 13] MARY G., MEZDOUR S., DELAPLACE G. *et al.*, "Modelling of the continuous foaming operation by dimensional analysis", *Chemical Engineering Research and Design*, vol. 91, pp. 2579–2586, 2013.

[MCC 93] MCCABE W.L., SMITH J.C., HARRIOTT P., *Unit Operations of Chemical Engineering*, 5th ed., McGraw-Hill Inc., Columbus, 1993.

[MET 57] METZNER A.B., OTTO R.E., "Agitation of non-Newtonian fluids", *AIChE Journal*, vol. 3, pp. 3–10, 1957.

[MID 81] MIDOUX N., Maquettes et similitudes, study course, École Nationale Supérieure des Industries Chimiques, Nancy, 1981.

[PAI 05] PAINMANAKUL P., LOUBIÈRE K., HÉBRARD G. *et al.*, "Effect of surfactants on liquid-side mass transfer coefficients", *Chemical Engineering Science*, vol. 60, pp. 6480–6491, 2005.

[PAU 04] PAUL E.L., ATIEMO-OBENG V.A., KRESTA S.M., *Handbook of Industrial Mixing: Science and Practice*, John Wiley & Sons, 2004.

[PAW 91] PAWLOWSKI J., *Veränderliche Stoffgrössen in der Ähnlichkeitstheorie*, Salle & Sauerländer, 1991.

[PET 11] PETIT J., HERBIG A.L., MOREAU A. *et al.*, "Influence of calcium on β-lactoglobulin denaturation kinetics: implications in unfolding and aggregation mechanisms", *Journal of Dairy Science*, vol. 94, pp. 5794–5810, 2011.

[PET 12] PETIT J., HERBIG A.L., MOREAU A. *et al.*, "Granulomorphometry: a suitable tool for identifying hydrophobic and disulfide bonds in β-lactoglobulin aggregates. Application to the study of β-lactoglobulin aggregation mechanism between 70 and 95°C", *Journal of Dairy Science*, vol. 95, pp. 4188–4202, 2012.

[PET 13] PETIT J., SIX T., MOREAU A. *et al.*, "β-lactoglobulin denaturation, aggregation, and fouling in a plate heat-exchanger: pilot-scale experiments and dimensional analysis", *Chemical Engineering Science*, vol. 101, pp. 432–450, 2013.

[QUA 95] QUARINI G.L.,"Thermalhydraulic aspects of the ohmic heating process", *Journal of Food Engineering*, vol. 24, pp. 561–574, 1995.

[RIC 13] RICHARD B., LE PAGE J.F., SCHUCK P. *et al.*, "Towards a better control of dairy powder rehydration process", *International Dairy Journal*, vol. 31, pp. 18–28, 2013.

[SAI 62] SAINT-GUILHEM R., *Les principes de l'analyse dimensionnelle, invariance des relations vectorielles dans certains groupes d'affinités*, Gauthier-Villars, 1962.

[SIE 36] SIEDER E.N., TATE G.E., "Heat transfer and pressure drop of liquids in tubes", *Journal of Industrial and Engineering Chemistry*, vol. 28, pp. 1429–1435, 1936.

[SZI 07] SZIRTES T., *Applied Dimensional Analysis and Modelling*, 2nd ed., Butterworth-Heinemann, Oxford, Waltham, 2007.

[TAY 32] TAYLOR G.I., "The viscosity of a fluid containing small drops of another fluid", *Proceedings of the Royal Society*, London, vol. 138, pp. 41–48, 1932

[VAN 79] VAN'T RIET K., "Review of measuring methods and results in non-viscous gas-liquid mass transfer in stirred vessels", *International and Engineering Chemistry Process Design and Development*, vol. 18, pp. 357–364, 1979.

[VIS 97] VISSER J., JEURNINK T.J.M., "Fouling of heat exchangers in the dairy industry", *Experimental Thermal and Fluid Science*, vol. 14, pp. 407–424, 1997.

[XUE 06] XUEREB C., POUX M., BERTRAND J., *Agitation et mélange: Aspects fondamentaux et applications industrielles*, L'Usine Nouvelle, Dunod, 2006.

[ZHO 98] ZHOU G., KRESTA S.M., "Correlation of mean drop size with the turbulence energy dissipation and the flow in an agitated tank", *Chemical Engineering Science*, vol. 53, pp. 2063–2079, 1998.

[ZLO 79] ZLOKARNIK M., "Sorption characteristics for gas–liquid contacting in mixing vessels", *Advances in Biochemical Engineering/Biotechnology*, vol. 8, pp. 133–157, 1979.

[ZLO 02] ZLOKARNIK M., *Scale-up in Chemical Engineering*, 1st ed., Wiley VCH, Weinheim, 2002.

[ZLO 06] ZLOKARNIK M., *Scale-up in Chemical Engineering*, 2nd ed., Wiley VCH, Weinheim, 2006.

[ZWI 58] ZWIETERING T.N., "Suspending of particles in liquid by agitators", *Chemical Engineering Science*, vol. 8, pp. 244–253, 1958.

Index

Printed in the United States
By Bookmasters